The Institute of Biology's
Studies in Biology no. 27

The Membranes of Animal Cells

Second Edition

A. P. M. Lockwood
M.A., Ph.D., F.I.Biol.

Reader in Biological Oceanography,
University of Southampton

Edward Arnold

First published 1971
by Edward Arnold (Publishers) Ltd.
41 Bedford Square, London WC1B 3DQ

Reprinted 1972
Reprinted 1974
Second edition 1978

'. . . *the whole of intra-cellular biology is a matter of membranes* . . .' J. D. Bernal 1965

British Library Cataloguing in Publication Data

Lockwood, Antony Peter Murray
The membranes of animal cells. 2nd ed. – (Institute of Biology. Studies in biology; no. 27 ISSN 0537-9024)
1. Cell membranes
I. Title II. Series
591.8'75 QH601

ISBN 0-7131-2731-7

Printed and bound in Great Britain at
The Camelot Press Ltd, Southampton

General Preface to the Series

Because it is no longer possible for one textbook to cover the whole field of biology while remaining sufficiently up to date the Institute of Biology has sponsored this series so that teachers and students can learn about significant developments. The enthusiastic acceptance of 'Studies in Biology' shows that the books are providing authoritative views of biological topics.

The features of the series include the attention given to methods, the selected list of books for further reading and, wherever possible, suggestions for practical work.

Readers' comments will be welcomed by the Education Officer of the Institute.

1978

Institute of Biology
41 Queen's Gate
London SW7 5HU

Preface to the First Edition

Unfortunately for the student, the literature on membranes is already voluminous and not without its conflicts of opinion and duplication of terminology. The primary function of this booklet, therefore, is to provide an elementary account for the beginner shorn of at least some of the 'ifs' and 'buts'. Over-simplification will be obvious to the specialists but the author will have achieved his aim if the student is stimulated to extend his reading to more detailed accounts of the topic.

I am much indebted to Dr F. S. Billett for reading the draft of this booklet and for pointing out various errors.

Southampton, 1971 A. P. M. L.

Preface to the Second Edition

There is a calculation in Lull's *Organic Evolution* to the effect that if all the progeny of a single *Paramecium* were to survive and divide in their turn then, after 9 000 generations (fifteen years) the mass of organisms 'would exceed the confines of the known universe and the rate of growth would be extending the circumference into space with the velocity of light'. The literature expansion on the topic of membranes in the seven years since this booklet was first published suggests that an almost equally daunting explosive growth is in progress. The extent of this research interest is to be welcomed since it reflects the recognition of the importance of membranes in almost every facet of cell function though it does have the unfortunate corollary that still greater selection and over-simplification is necessary in a basic summary text such as this.

Southampton, 1978 A. P. M. L.

Contents

1 Introduction

Biological theories, like sartorial fashions, not infrequently turn full circle. A hundred years ago the distinguished microscopist C. G. Ehrenberg claimed to be able to detect a complex series of internal organs within the cells of protozoa but his ideas were derided. The last thirty years have shown, however, that, though his views were over-extravagant, he was at least correct in supposing that cells have a highly organized internal structure composed of membrane-bounded vesicles.

Interest in these intracellular membranes has become increasingly intense with the realization that they do not just play a passive role in segregating different regions of the cell but that their functions embrace every facet of cell activity.

The multiple nature of their metabolic involvement and the complexity of their structure have made membranes the natural meeting point of the sciences with electron microscopists, physical chemists, biochemists and biophysicists all approaching them from different viewpoints. The results obtained by this concerted study have made it obvious that knowledge of the precise structure and functioning of different membranes will open the way not only to an understanding of what constitutes life at the molecular level but also to that vital selective control of cell and tissue function necessary in the treatment of cell malfunction and tissue transplants. The biologist therefore neglects a study of membranes at his peril.

The full complexity of the various roles of membranes has only recently been recognized but already it seems perhaps not too far reaching to suggest that the subject of membranes will prove to be as important a theme in the knowledge of how cells function as the concept of evolution has been to biology in general or as DNA has been to genetics.

Hundreds of chemical reactions and transfer processes are occurring continuously in every active cell, and membranes play a vital role in controlling these processes, in separating incompatible substances and in transporting materials about the cell. Some of the more complex chemical sequences are expedited because it appears that the enzymes which catalyse them are so arranged on membranes that the reactants can move readily from one to the next.

In a packaging role, intracellular membranes separate components of the cell which would be self-destructive if allowed to mix freely in the cytoplasm, they conserve and maintain regions of local concentration, regulate the passage of inorganic ions and complexes between compartments and provide the principal means for the trammelling, ordering and regulation of the metabolic processes which constitute life.

2 What and Where are the Membranes?

The advent of the electron microscope has greatly extended knowledge of intracellular membranes. Forty years ago the only membrane systems recognized with any degree of certainty were the *plasma membrane* separating the cytoplasm from the medium at the cell boundary, the *nuclear membrane* separating the nucleoplasm from the cytoplasm and an enigmatic system of vesicles, the *Golgi apparatus*, about whose reality controversy was rife. Various other cellular inclusions such as mitochondria, vacuoles and mitotic figures were known but that they were membranous structures remained to be established. Since then the number of intracellular organelles known to be associated with membranes has grown considerably. The most commonly represented

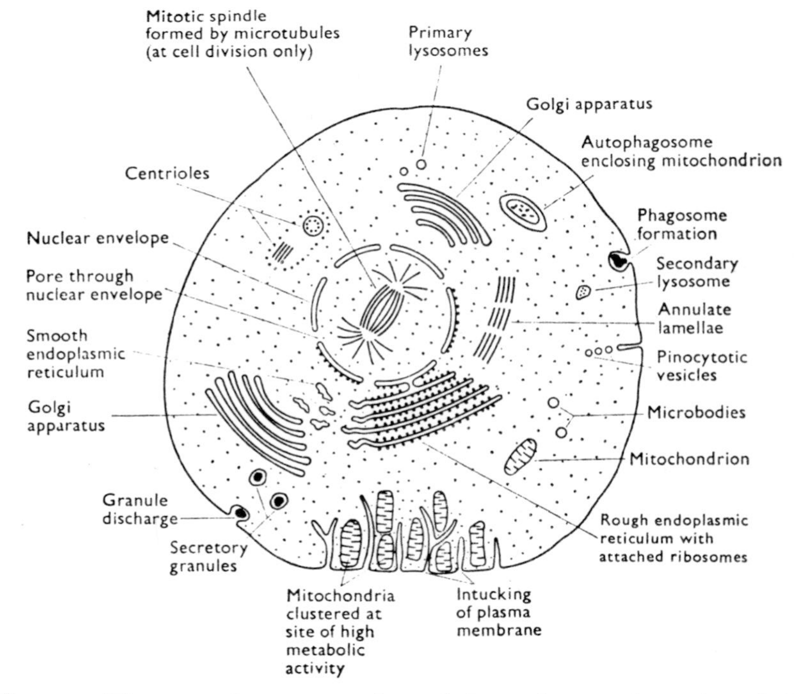

Fig. 2–1 Diagrammatic representation of the major membrane-bounded structures found in cells. N.B. Few, if any, cells will contain all these organelles simultaneously.

structures are listed in Table 1 and illustrated diagrammatically in Fig. 2–1. Their appearance in the electron microscope can be seen by reference to Figs 2–2, 2–3 and 2–4.

Not every type of cell has a full complement of the organelles outlined in Table 1 and the degree to which they are represented in different tissues

Table 1 Cell organelles and their primary functions

Plasma membrane	Diffusion barrier, active transport of sodium outwards, electrically excitable membrane. Principal enzymes include Na–K–Mg ATPase, glucose-6-phosphatase
Rough endoplasmic reticulum	Synthesis of protein
Smooth endoplasmic reticulum	Steroid metabolism, transport of materials from rough endoplasmic reticulum to Golgi apparatus
Golgi apparatus	Storage vesicles; 'packaging' of enzymes to be used elsewhere
Lysosomes	Vesicles containing enzymes used in autodigestion and breakdown of materials brought in by phagosomes and pinocytosis
Phagosomes	Vesicles, containing particulate material from outside the cell; formed by inversion of portions of the plasma membrane
Pinocytotic vesicles	Microscopic vesicles, formed by plasma membrane inversion, containing materials adsorbed onto the membrane
Multivesicular vesicles	Bodies containing smooth membranes. These belong to the lysosome family of organelles
Microtubules	Fine cylindrical structures which subserve a variety of roles in the cell
Microsomes	The name given to small fragments of cytomembrane, principally those of the endoplasmic reticulum and Golgi apparatus, produced as an artefact of homogenization and separated by centrifugation
Mitochondria	Membranous organelles responsible for the production of most of the energy donor substance, ATP, formed by cells
Microbodies	Vesicles containing the enzyme catalase and often involved in uric acid metabolism
Nuclear envelope	A two-layered membranous diffusion barrier between the cytoplasm and the nucleoplasm
Centrioles	A pair of self-replicating cylindrical bodies containing 11 parallel fibrils. They function in organizing the poles of the mitotic figure during cell division
Basal bodies	Structurally similar to centrioles, associated with the base of flagellae
Annulate lamellae	Present primarily in oocytes, function unclear
Microfibrils	Filaments, usually contractile, which are involved in aspects of cell or organelle movement, e.g. constriction of the cleavage furrow of dividing cells

also varies widely. Thus the endoplasmic reticulum is particularly extensive in cells, e.g. those of the pancreas, which manufacture and secrete protein; lysosomes are well developed in macrophages, which ingest particulate matter; pinocytotic vesicles are common in liver; annulate lamellae are characteristic of egg cells; Golgi vesicles are

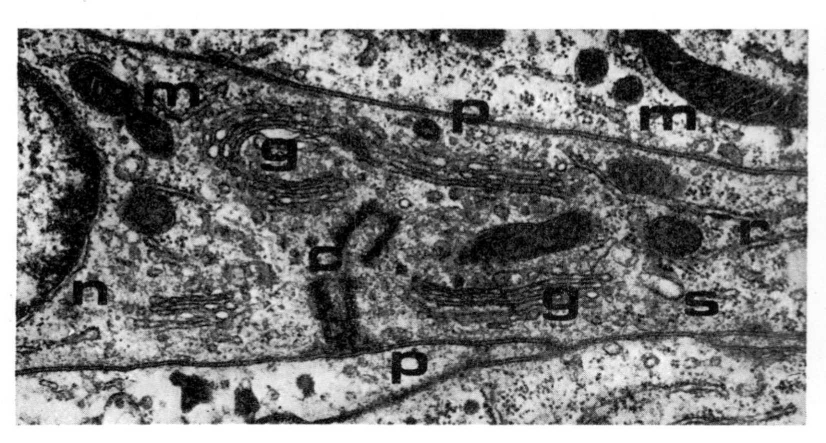

Fig. 2–2 Portion of an avian parathyroid gland parenchymal cell showing the principal organelles. c, centrioles; g, Golgi apparatus; m, mitochondrion; n, nuclear membrane; p, plasma membrane; r, rough endoplasmic reticulum; s, smooth endoplasmic reticulum. (× 1700.) (Reproduced by courtesy of Mr R. P. Gould.)

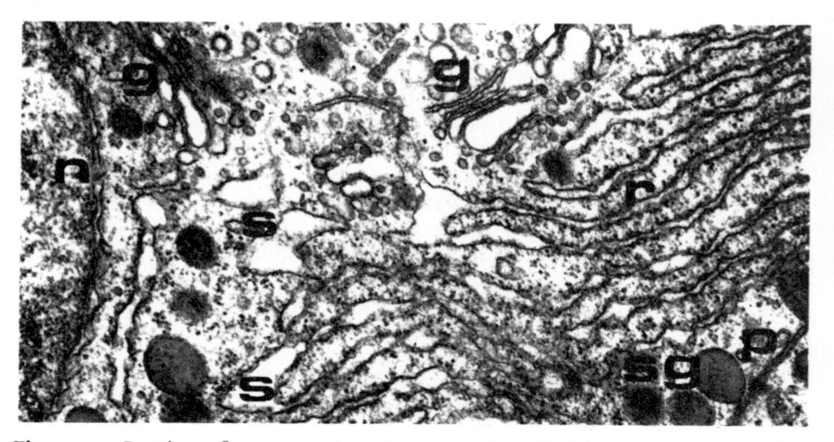

Fig. 2–3 Portion of a rat anterior pituitary cell. g, Golgi apparatus; n, nuclear membrane; p, plasma membrane; r, rough endoplasmic reticulum; s, smooth endoplasmic reticulum; sg, secretory granule. (× 28 200.) (Reproduced by courtesy of Mr R. P. Gould.)

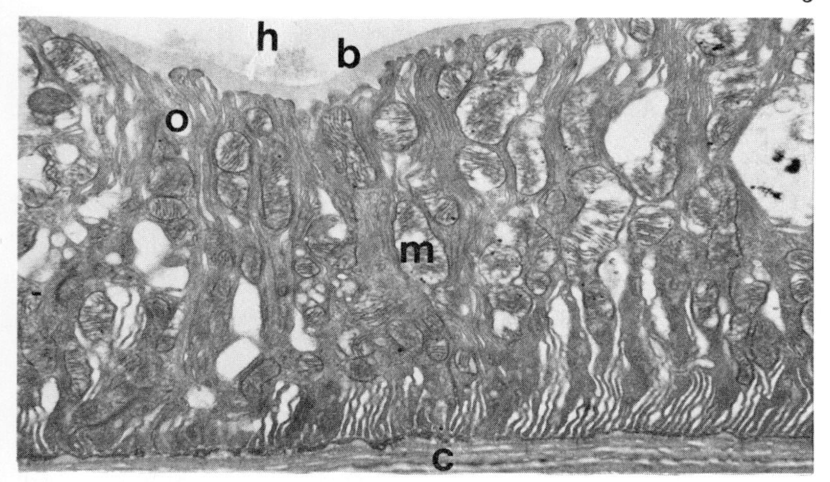

Fig. 2–4 Part of an ion-transporting epithelial cell of the gill of *Gammarus duebeni* (Amphipoda) showing the typical intuckings of the apical plasma membrane in a transporting cell. o, plasma membrane intuckings; m, mitochondrion; c, cuticle at external surface of the gill; b, basement membrane; h, haemolymph (blood). (× 6000).

enlarged in storage tissues and mitochondria concentrate in regions of high energy expenditure in the cell.

One of the primary functions of membranes is to provide a surface for the location of enzymes. The relatively enormous area which can be available for this purpose is clearly illustrated by the estimate of the size and number of the major organelles of a liver cell given in Table 2. To put the figures in perspective one should note that a cube of 100 mm side of such cells (giving a volume some 3 times less than that of a human liver) would contain over 9000 m^2 of endoplasmic reticulum. This is approximately equivalent to the area of 32 tennis courts.

Table 2 Organelles of a single rat liver cell. (From WEINER, J., LOUD, A. V., KIMBERG, D. V. and SPIRO, D. J. (1968). *Cell Biol.*, 37.)

Volume μm^3	
Total cytoplasm of cell	5 100
Mitochondria (total)	995
Lysosomes (total)	10
Membrane area μm^2	
Smooth endoplasmic reticulum	17 000
Rough endoplasmic reticulum	30 400
Mitochondrial outer membrane	7 470
Mitochondrial inner membrane	39 600
Total number of mitochondria 1160	

3 Membrane Composition and Configuration

3.1 Configuration

Anyone who has had the dubious pleasure of washing the greasy plates after Sunday lunch will be well aware of the two facts that meat contains fat and that fat is not readily miscible with water, especially cold water. This immiscibility of lipid and water is put to good use in cellular and intracellular membranes, most of which owe their low permeability to water-soluble materials to the high proportion of lipids in their composition.

Isolated plasma membranes of erythrocytes (red blood cells) and many other cells contain about 40% lipid, 0–10% carbohydrate and 50–60% protein. The manner in which the relative impermeability of cell membranes derives from their lipid content may be understood by consideration of what happens at the molecular level when lipid and water interact. Fatty acids [e.g. compounds with the general structure ($CH_3(CH_2)_nCOOH$)] have a polar end (—COOH) which can interact with water molecules and a non-polar, strongly hydrophobic, group (CH_3—) at the other end of the hydrocarbon chain (see p. 7). In consequence fatty acid molecules placed at an air-water interface tend to orientate themselves in a layer one molecule thick with the polar groups in contact with the water and the non-polar hydrocarbon chains sticking up into the air. If just sufficient fatty-acid molecules are present to cover the water surface a close-packed array of molecules is formed (Fig. 3–1a) which has a high measure of impermeability to hydrophilic substances.*

When more molecules are present than are required to form a monolayer a bimolecular leaflet of lipid may be produced (Fig. 3–1b).

Early measurements of the amount of lipid present in the cell membrane of red blood cells showed that there was just sufficient present to cover the surface area of the cell with such a bimolecular leaflet and it was therefore suggested that the cell membrane consists of a lipid bilayer which is coated on each side by proteins (Fig. 3–2). This hypothesis, due to H. Davson and J. F. Danielli, at first proved most attractive since it accounted for a number of properties of cell membranes including their dimensions and appearance in electron micrographs, the presence of lipid and protein and the fact that lipid soluble substances penetrate membranes preferentially. However, studies over the last few years have

* This property has, incidentally, been utilized in reservoirs in the warmer parts of the world to slow the loss of water by evaporation. For example, experiments in which a monolayer of cetyl alcohol was spread on Lake Umberumberka in Australia reduced evaporative water loss by up to 50%.

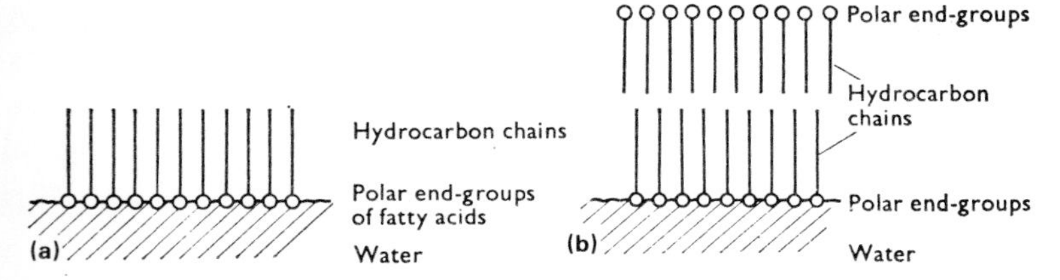

Fig. 3–1 Interaction between lipid molecules and water. (**a**) Close-packed molecules orientated at an air-water interface with the polar (hydrophil) groups in contact with the water and the non-polar (hydrofuge) hydrocarbon chains sticking up into the air. (**b**) A 'doublet' of lipid molecules is formed when too much lipid is present to be contained by a monolayer.

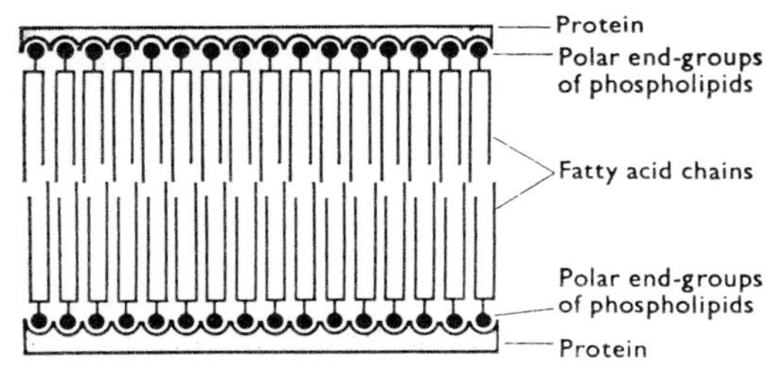

Fig. 3–2 Stylized drawing of the molecular organization of a membrane according to the 'unit membrane' theory. A lipid bilayer is coated on either side by protein layers which interact with the polar head-groups of phospholipid.

suggested that, though the concept of the 'unit membrane' based on a ground structure of a bimolecular leaflet of lipid can still be accepted in broad terms, the original Davson–Danielli concept as regards the location of the proteins must be revised. It now appears that the proteins are actually incorporated in the body of the membrane and may even, in some cases, be so located through the membrane that they project on either side (Fig. 3–3). Evidence for this conclusion is based on observations which suggest that:

(1) membrane proteins are essentially globular rather than having the expanded structure to be expected if they were superficially located;

(2) some protein molecules can be tracer labelled, or experimentally attacked by enzymes from both sides of the membrane whilst others (those thought to be restricted to a single surface of the membrane) can only be labelled from one side;

(3) globular structures of similar dimensions to membrane proteins can be seen in the interstices within the lipid bilayers when membranes are frozen and then fractured in the horizontal plane.

The finding that some proteins are quite firmly bound to lipids and that certain enzymes, including glucose-6-phosphatase and Na–K–Mg

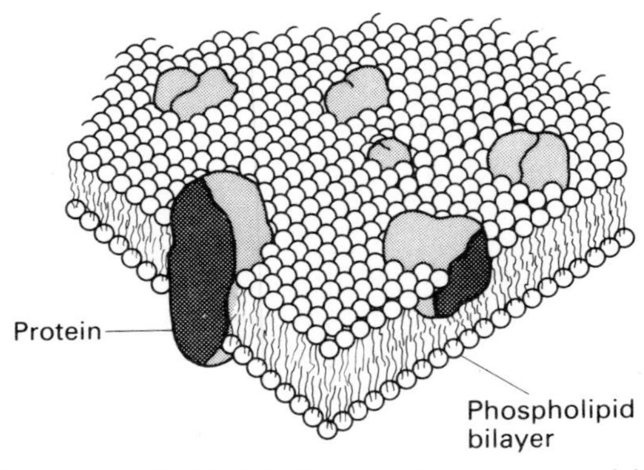

Fig. 3–3 The current lipid-globular protein mosaic model of the plasma membrane. Lipid molecules are shown as hollow circles with the hydrocarbon tails projecting into the body of the membrane; proteins are the larger masses incorporated into the membrane or projecting right through. The upper face, as shown, will be that in contact with the cytoplasm. The carbohydrate elements of the glycoproteins which project from the external surface are not visible in this diagram. (After SINGER, S. J. and NICHOLSON, G. L. (1972) *Science*, **175**, 720–31.)

ATPase will not function in the absence of lipid also suggests that the proteins are constituent components of the membrane.

Membranes should not be regarded as rigid, static structures. The overall viscosity, as pointed out by G. L. Nicholson, is comparable with that of light machine oil so that the general consistency is fluid. Considerable mobility of the molecules in the plane of the membrane is therefore possible though it appears that there is essentially an 'upstairs downstairs' separation of the two lipid bilayers in so far as 'flip flop' interchange of a phospholipid molecule from one layer to the other is a relatively rare event by comparison with alteration of position within a layer. The ability of molecules to move laterally and change their associations with one another makes it possible for membranes to become mosaics of different structural and functional regions in both space and time. Both such associations of molecules form specialized *domains* within membranes and the variety of molecules composing membranes determine the configuration and function.

3.2 Chemical composition

Lipids and proteins together account for almost the entire composition of membranes; but each of these two general types of compound occur in a wide variety of forms.

3.2.1 *Lipids*

Two general classes of lipid occur in organisms: (1) simple lipids, (2) compound lipids.

Simple lipids are composed of glycerol and fatty acids and have the generalized formula shown below.

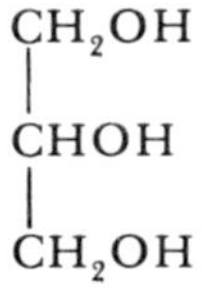

$$\begin{array}{ll} CH_2OH & CH_2O{-}R_1 \\ | & | \\ CHOH & CHO{-}R_2 \\ | & | \\ CH_2OH & CH_2O{-}R_3 \\ \text{Glycerol} & \text{Triglyceride} \\ & \text{(simple lipid)} \end{array}$$

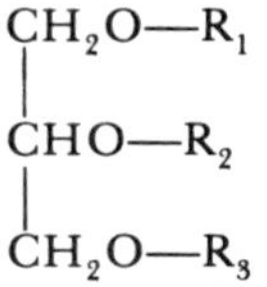

where R_1, R_2 and R_3 are fatty acids.

Simple lipids are important in fat deposits but, though present in muscle plasma membranes, they do not normally contribute appreciably to cellular membranes.

The fatty acids combined in both the simple and compound lipids have a variety of chain lengths. Three of the most commonly occurring forms are palmitic, stearic and oleic acids.

Palmitic acid	$CH_3(CH_2)_{14}COOH$
Stearic acid	$CH_3(CH_2)_{16}COOH$
Oleic acid	$CH_3(CH_2)_7CH{=}CH(CH_2)_7COOH$

Fatty acids which, like oleic acid, contain double bonds are termed 'unsaturated'. Those with no double bonds, as in palmitic and stearic, are 'saturated'.

Compound lipids are more complex in composition than the simple lipids and, in addition to fatty acids, may also include glycerol, or similar compounds and nitrogenous bases. Steroids are commonly classed with this group though they do not contain fatty acids. The three most important groups of compound lipids are (1) glycerophosphatides, (2) sphingolipids and (3) steroids.

(1) Glycerophosphatides are a group of compound lipids based on simple lipids in which one of the fatty acids is substituted. Several types occur:

(a) phosphatidyl choline (lecithin) composed of glycerol, two fatty acids, phosphate and choline;
(b) phosphatidyl ethanolamine composed of glycerol, two fatty acids, phosphate and ethanolamine;
(c) phosphatidyl serine composed of glycerol, two fatty acids, phosphate and serine;
(d) phosphatidyl inositol composed of glycerol, two fatty acids, phosphate and inositol.

These last three compounds are sometimes referred to as *cephalins* though this term is more usually restricted to phosphatidyl ethanolamine and phosphatidyl serine alone.

The virtual absence of lipid-bound amino groups on the outward facing surface of the membrane suggests that phosphatidyl choline (lecithin) is the dominant lipid of the outer bilayer whilst the amino-acid lipids (phosphatidyl serine and phosphatidyl ethanolamine) are the principal phospholipids of the inner bilayer. Glycolipids are thought to be restricted to the outer layer.

The amino-acid phospholipids are usually present in a somewhat lower concentration than the sum of lecithin and the glycolipids which is compatible with the suggested distribution since the inner bilayer would be expected to have a lower lipid component than the outer so that the layer may incorporate the associated proteins (cf. Fig. 3–3).

Membranes often contain small proportions of materials which, though otherwise comparable to the four listed above, give aldehyde reactions. These are known respectively as phosphatidal choline, phosphatidal ethanolamine, phosphatidal serine and phosphatidal inositol.

A more complex glycerophosphatide, cardiolipin, isolated from heart membranes, contains two glycerols, four fatty acids and two phosphate groups.

(2) Sphingolipids differ from the glycerophosphatides in lacking glycerol: instead they contain the nitrogenous base sphingosine. The two major compounds of the group are (a) *sphingomyelin* (sphingosine, one fatty acid, phosphate and choline) and (b) *cerebroside* (sphingosine, one fatty acid and galactose).

(3) Cholesterol. This steroid is a major component of plasma membranes but only occurs to a minor extent in intracellular membranes. Most of the cholesterol in plasma membranes is associated with the outer bilayer.

The approximate steric structure of these compound lipids is illustrated in Fig. 3–4.

Minor components of membranes include sialic acid, phosphatidic acid, glycolipids and inorganic ions, particularly calcium, but, by comparison with the lipids mentioned above and proteins, they occur in only trivial amounts.

Although compound lipids contribute to the membranes of every cell they are by no means constant in amount in different tissues or membrane systems or indeed even in one type of plasma membrane under different environmental circumstances. Such variation reflects membrane specialization and the need to maintain a specific viscosity in order to function normally.

The degree of mobility of lipid molecules is dependent upon whether their *phase transition temperature* (effectively the melting point) is above or

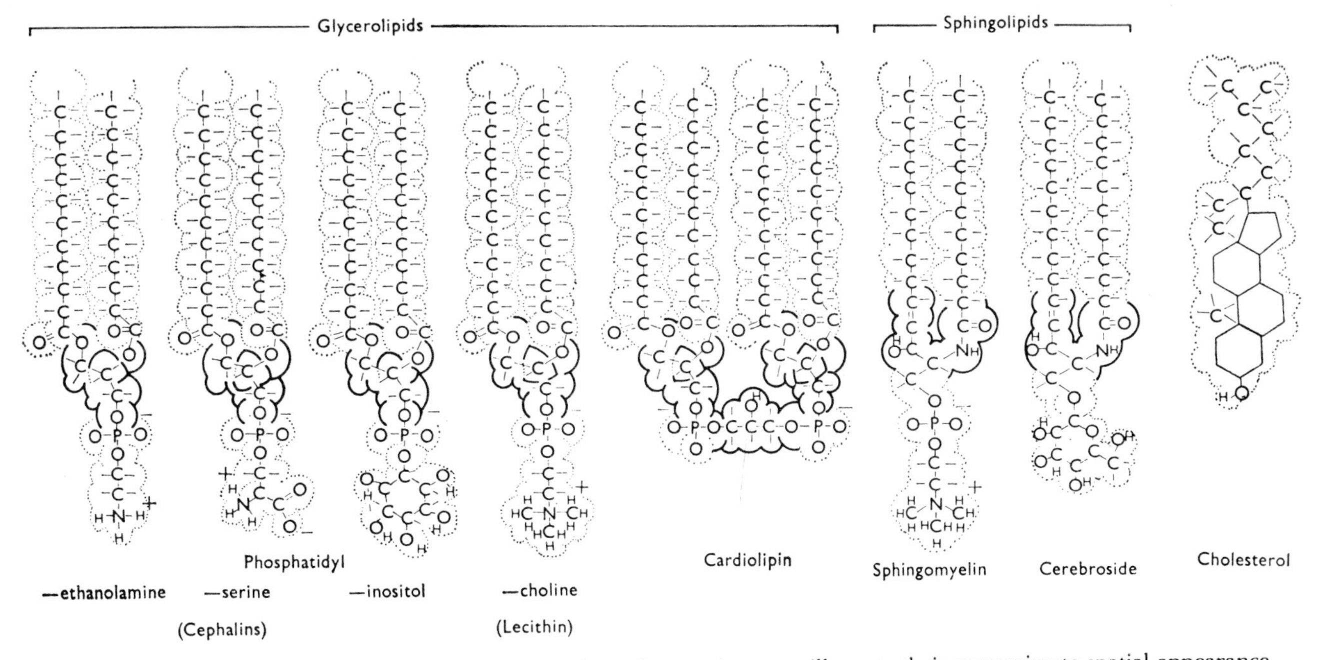

Fig. 3–4 The chemical formulae of the major lipids of membranes drawn to illustrate their approximate spatial appearance. (From FINEAN, J. B. (1961). In *Chemical Ultrastructure in Living Tissue*. Charles C. Thomas, Illinois.)

below the ambient temperature. The phase transition temperature is determined by the chain length and degree of saturation of the molecule. The shorter the chain length and the greater the degree of unsaturation, the lower is the melting point. In addition the phase transition can be influenced by the presence of cholesterol. Cholesterol serves to blur the distinction between the solid ('frozen') and liquid ('melted') phases since it has the effect of decreasing molecular mobility in the liquid phase and increasing it in the solid phase. The appropriate degree of mobility is important to normal functioning of membranes and there is evidence from both dietary and temperature studies of the dynamic regulation of this property. Thus, although unusual dietary lipid intake can result in variation of the lipid composition of membranes there is a tendency for a balance to be maintained between the saturated and unsaturated components such that a more or less constant overall viscosity is maintained. Similarly, response to temperature change can involve adjustments within membranes. When the protozoan *Tetrahymena*, cultured at 28°C, are suddenly exposed to 10°C there is some degree of gelling (i.e. freezing) of the plasma membrane. Within 24 h, however, changes in the membrane components lower the overall freezing point restoring the fluid nature of the membrane.

Despite such control of average viscosity, segregation of particular lipid types can create zones within a membrane of a greater or lesser mobility than the mean. There are also considerable variations in the composition of different membranes confirming the proposition that there must be a variety of ways in which membranes can be assembled from component molecules and that differences in composition can be related to membrane function. Myelin (from nerve sheaths), chloroplast membranes, mitochondrial membranes and red cell plasma membranes illustrate these differences with respect to their content of glycerophosphatides and sphingolipids. The glycerophosphatide content of mitochondrial membranes is high (over 75% of the total lipid); it is smaller in myelin and red cell ghosts (about 33% of the lipid) and still less in chloroplast membranes (12%). Sphingolipids are high in myelin (25%), less important in red cell ghosts (15%), and still lower in mitochondria.

Not only does the total content of phospholipid vary but the fatty acids they incorporate also differ from membrane to membrane. The lipids of myelin are rich in saturated long chain fatty acids admixed with those of medium chain length; medium chain length unsaturated fatty acids are most frequent in red cell ghosts and mitochondrial membranes are also rich in unsaturated fatty acids. That such differences in composition reflect the functional role of a membrane, or of specialized regions within a membrane, is indicated by experimental attempts to reconstitute the oxidative phosphorylation system after disruption of mitochondrial membranes. Success is dependent upon the relative proportions of phosphatidyl ethanolamine and lecithin present.

In addition to variations in composition between the membranes of different tissues there are also variations between similar membranes taken from a range of species. The red cell ghosts of mammals show a steady increase in the content of oleic acid at the expense primarily of palmitic acid in the series rat/man/rabbit/pig/horse/ox/sheep (Table 3). Hand in hand with these differences in composition go changes in physiological properties. As the oleate content increases from rat to sheep so the permeability of the red cell to urea, thio-urea and glycerol declines.

Table 3 Some major fatty acids of red blood cell ghosts of a variety of species of mammal (mol. %). (Extract from KOGL, F., DE GIER, J., MULDER, L. and VAN DEENEN, L. L. M. (1960). *Biochim. Biophys. Acta,* 43.)

	Rat	*Man*	*Rabbit*	*Pig*	*Horse*	*Ox*	*Sheep*
Palmitate	44	37	36	30	21	13	12
Stearate	22	15	11	14	14	14	7
Oleate	18	26	25	35	30	52	61
Linoleate	15	17	23	17	29	15	10

Variations between species are also shown at the level of the compound lipids (Table 4). In this case there is an interesting correlation between the levels of the various phospholipids and the sensitivity to snake bite. The venom of many snakes and in particular that of the cobras, rattlesnakes and krait, owes much of its virulence to the presence of phospholipase-a. This enzyme breaks down lecithins and cephalins and so, by destruction of the plasma membrane, causes lysis of cells (particularly in red cells). Sphingomyelin is not affected by phospholipase-a and cells with a high proportion of this lipid in their plasma membranes are less likely to suffer lysis by low concentrations of the enzyme than are cells with a high lecithin content. This may perhaps explain the low sensitivity of sheep and goats to snake bite by comparison with that of other mammals.

Table 4 Phospholipids (as % phospholipid phosphate) in red cell membranes from different species. (From DITTMER, J. C. (1962), p. 238 in *Comp. Biochem.*, III. Edited by M. FLORKIN.)

	Lecithin	*Cephalin*	*Sphingomyelin*
Man	29·2	14·4	17·6
Goat	9·8	6·8	34·4
Sheep	7·7	5·4	42·5

It is entertaining to speculate that the high sphingomyelin and low lecithin content of sheep and goat cell membranes might indeed be due to natural selection with respect to snake bite since their mode of life renders these ungulates more liable to this hazard than many other mammals.

Phospholipase-a is also present in a wide variety of other toxins ranging from the sting of wasps, bees and scorpions to the bite of some spiders. Another phospholipase, phospholipase-c, is the principal toxin of *Clostridium welchii*, the bacterium responsible for one form of gas gangrene, a disease characterized by cell lysis.

The steroid, cholesterol, commonly accounts for a large part (approximately 35%) of the total lipid content of plasma membranes. It makes a much smaller contribution to the composition of intracellular membranes and is absent from the inner membranes of mitochondria and the membranes of most, though not all, bacteria.

3.2.2 Proteins

Much of the protein associated with membranes is accounted for by bound enzymes but it is possible that at least some is involved in purely structural and non-catalytic roles.

In the red blood cell membrane about twenty proteins and glycoproteins (combinations of sugars and protein) are recognized as forming the bulk of the protein component. Of these about a quarter can be separated from other membrane components fairly readily and it is thought that this fraction (containing largely tectin-a and actin) probably represents molecules which are located peripherally on the inner membrane surface. Few, if any, proteins are exclusively associated with the outer lipid bilayer, several do, however, span the membrane. In this latter category two of the major contributors are Component *a* (=Component III) and glycophorin.

Component *a*, a protein with a molecular weight in the range 90 000–100 000, is thought to be involved in anion translocation across the plasma membrane. Glycophorin, a sialoglycoprotein containing some 60% carbohydrate by comparison with the 6–8% in Component III, bears the ABO and MN antigens and consequently is intimately involved in cell-recognition processes.

Although proteins spanning membranes tend not to differ strikingly from other proteins in respect of their overall amino-acid composition their location indicates that they must be amphipathic (i.e. have regions which are either hydophobic or hydrophilic according to whether the part is positioned within the membrane or at one of the surfaces). Hydrophobic regions often react strongly with lipids and sometimes are tightly bound to phospholipid elements in the membrane. This binding, together perhaps with inter-connection of carbohydrate groups on the glycoproteins, serves to ensure the positioning and maintenance of such proteins in their transmembrane location.

Glycoproteins are virtually absent from membrane systems in the cytoplasm. In the plasma membrane they are always so orientated that the carbohydrate moiety is peripheral to the outer lipid bilayer. The carbohydrate is generally in the form of short lengths, *c.* 10 residues each, which are attached to about 10 amino acids close to the amino terminal group of the protein backbone. The carbohydrates thus 'bristle' from the cell surface; since they bear negative charge, they may assist in keeping adjacent cells apart (Fig. 4–3).

3.2.2.1 Enzymes. Proteins have a three-dimensional structure and enzymic function depends on the twin factors of maintenance of the correct molecular configuration and the exposure of the catalytically active regions. The incorporation of enzymic proteins in membranes provides the means by which the appropriate steric arrangement of the molecules can be sustained and also the spatial separation of enzymes which might interact with each other. Attachment of enzymes to membranes also has the advantage that the separate enzymes mediating the various steps in a complex series of reactions, such as the electron transfer process of the cytochrome chain, can be located together in such a manner that the chemical conversions can proceed with the maximum expedition.

Probably all plasma and cytomembranes have associated enzymes. Certain enzymes are characteristic of particular membranes though not necessarily restricted to that one type. Thus, Na–K–Mg ATPase is typical of the plasma membrane; cathepsins and phosphoprotein phosphatase of lysosomes; cytochrome-c oxidase and succinic-cytochrome-c reductase of mitochondrial cristae; fatty acid elongating enzymes of mitochondrial outer membranes; peroxisome enzymes of microbodies and glycoprotein-glycosyl transferases of smooth membranes.

Specialization has also resulted in differences in the enzyme components of the same membrane system from different tissues. Thus the plasma membrane of intestinal and kidney cells has higher levels of alkaline phosphatase than does that of liver cells. Also, as might be expected, the plasma membrane of intestinal cells is particularly well equipped with bound digestive enzymes which supplement the activity of enzymes released into the gut lumen.

In yeast cells too the digestive enzymes invertase and phosphatase are bound to the outside of the plasma membrane. Such binding of digestive enzymes to the outer surface of cell membranes has the obvious advantage, particularly in the case of microorganisms, that neither the products of digestion nor the enzymes are dispersed in the body of the medium.

4 A Survey of Cellular Membranes

4.1 Plasma membrane (Figs 2–2 and 4–1)

The plasma membrane is perhaps the most variable of all the membrane systems of cells in form, chemical composition and properties. This is not surprising since, as the barrier between the cell and its environment, the plasma membrane must be adapted to meet the requirements of the various types of cell specialization. In addition to its role (discussed in Chapter 7) of maintaining concentration gradients between inter- and extracellular compartments, the cell membrane also has a diversity of functions in connection with cell recognition, intercellular communication, growth and control of cell division.

4.1.1 General form and composition

In electron micrographs the plasma membrane is approximately 7.5–9 nm thick and of relatively uniform appearance except where intercellular junctions (cf. section 4.1.2) interrupt the surface.

The gross configuration of the plasma membrane ranges from the relatively regular biconcave outline of mammalian red blood cells to the condition shown by the kidney tubule cells which are highly plicated on one surface and bear tiny finger-like processes on the opposite face (Fig. 4–2). The multiple infolding of the surface in the kidney cells increases the surface of the cell and thus enlarges the area over which exchange can occur. In higher organisms the cells, except for the motile blood cells, are generally stable in outline when mature though even here epithelial cells retain the ability to change shape and this faculty can be utilized during wound healing.

Generalizations about the chemical composition of plasma membranes are also made difficult by the degree of variability. Most of the cells which have been analyzed have plasma membranes composed of approximately 40% lipid and 60% protein. The myelin sheath of nerves (which is derived from the cell membrane of Schwann cells, p. 72) reverses this ratio, however, since it is 70% lipid and 30% protein. The major part of the lipid in the plasma membranes of liver and red blood cells is accounted for by cholesterol and phospholipid (Table 5) but in muscle cells triglycerides are present in higher concentration than the phospholipid.

To some extent plasma membranes differ chemically from the internal membranes of cells. Phospholipids and proteins are common to both but it appears that one group of compounds, glycolipids (combinations of

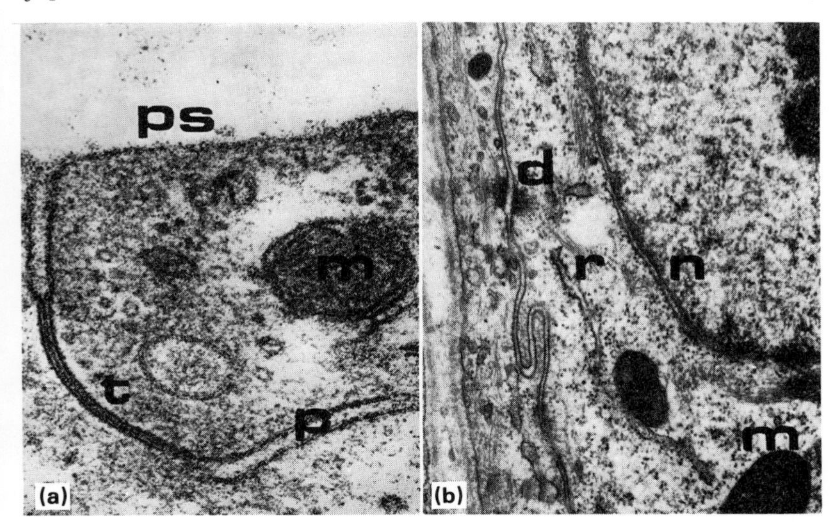

Fig. 4–1 (**a**) Tight junction between plasma membranes of neighbouring epithelioid cells, m, mitochondrion; p, normal spacing of cell membranes; ps, plasma membrane at surface of cell; t, tight junction. (× 30 700.) (Reproduced by courtesy of Mr R. P. Gould.) (**b**) Desmosome at plasma membrane between parathyroid parenchymal cells, d, desmosome; m, mitochondrion; n, nuclear membrane; r, rough endoplasmic reticulum. (× 81 800.) (Reproduced by courtesy of Mr R. P. Gould.)

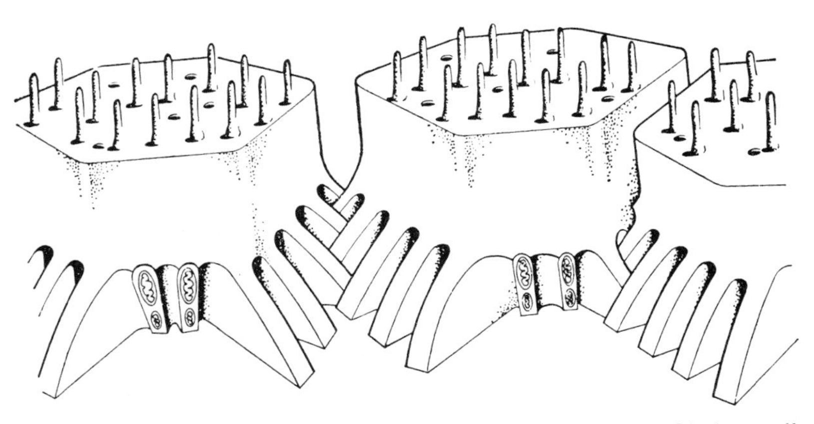

Fig. 4–2 Simplified reconstruction of convoluted tubule cells of kidney cells showing the interlocking fluted boundaries created by intucking of the plasma membrane. *In vivo* the cells are more variable than illustrated. (From RHODIN, J. (1958). *Int. Rev. Cytol.*, **7**.)

lipid and sugars) may be confined to the former and both cholesterol and sialic acid are present in markedly higher concentrations in the plasma membrane than in internal membranes.

As well as variations in composition between different tissues from a single species, differences also occur in membrane composition of the same tissue from different species. For example, the plasma membranes of the red blood corpuscles (RBCs) of cattle contain both sialic acid and hexosamine whilst horse, dog and cat RBCs contain sialic acid but little or no hexosamine; in sheep and goats precisely the opposite condition prevails, hexosamine being present and sialic acid in low concentration or absent.

Table 5 The lipid composition of rat liver plasma membrane. All values as percent total lipid. (From DOD, B. J. and GRAY, G. M. (1969). Table I in *Biophys. Biochem. Acta*, 150) (* Glycolipid by difference.)

Cholesterol	17	
Free fatty acid		
Cholesterol esters		
Triglycerides	17	
Unidentified		
		Phospholipids as % total phospholipid
Phosphatidyl ethanolamine		11
Phosphatidyl serine		6
Phosphatidyl inositol	59	6
Phosphatidyl choline		41
Cardiolipin		—
Sphingomyelin		33
Ceramide trihexoside		
Ceramide dihexoside		
Ceramide monohexoside	7*	
Gangliosides		

The proportions of various fatty acids present also show specific variations (see Table 3). Wide differences in the permeability of different cell membranes to water (the cell surface of *Amoeba* is about 100 times less permeable than the plasma membrane of a red blood cell) probably reflect such variations in lipid content.

Much of the protein associated with cell membranes is enzymic though, as one would expect, this component too shows some variety according to the tissue studied. Thus, membranes from the epithelial cells lining the gut lumen contain more alkaline phosphatase than do liver cell membranes and also a wide range of digestive enzymes not present in the latter. An enzyme that is characteristic of all plasma membranes is Na–K–Mg ATPase which plays an important part in ion transport, another links sugar to protein to make glycoproteins. Glycoprotein formed by this latter enzyme often coats the outer surface of plasma membranes to a thickness of 5–7 nm. One important class of these membrane-bound glycoproteins form the blood group factors. Others, released from the cell surface, are components of such divere systems as serum proteins, cartilage, synovial fluid, mucus and the vitreous fluid of the eye.

Plasma membranes influence the passage of everything that enters or leaves cells. To achieve this they must be relatively impermeable to small molecules but not so impermeable that the passage of the respiratory gases is restricted. That this end has been realized is indicated by the 100-fold greater permeability to oxygen shown by plasma membranes by comparison with simple monolayers of hexadecanol or octadecanol. Probably this degree of permeability is achieved by the incorporation of large, awkwardly shaped molecules, which make close packing of the phospholipid more difficult. Hence random thermal movement of the molecules composing the plasma membrane is likely to create temporary pores large enough to permit the passage of oxygen molecules.

Similar effects on membrane permeability may be responsible for the finding that animals which would normally be rendered unconscious by exposure to high hydrostatic pressure will recover consciousness if exposed to appropriate amounts of *narcotic* gases, such as N_2O. One plausible suggestion put forward to explain this effect, which is of potential interest in deep sea diving studies, is that pressure compression of the membrane in sensitive nerve cells is offset by expansion due to incorporation of the gas molecules so that the ion permeability and electrical properties of the membrane are restored to normal.

The question as to whether other diffusing substances such as urea and inorganic ions pass through the lipid layers during thermal displacement of the molecules or via more permanent aqueous pores is a matter which has aroused more than a little controversy. The problem is clearly an important one if we are to have adequate knowledge of membrane structure but unfortunately experimental results are conflicting. Theoretically, substances which have a molecular diameter of less than about 0.8 nm could pass through the space occupied by three hydrocarbon tails of phospholipids or one cholesterol molecule. Temporary displacement of lipid molecules creating a pore of this size could thus permit the passage of Na^+ (hydrated diameter 0.56 nm), K^+ (hydrated diameter 0.38 nm) and urea (0.2 nm) but would exclude the passage of larger ions such as Ca^{++} (0.96 nm) and Mg^{++} (1 nm). The actual rate of passage of substances across membranes is not dependent on size alone, however, and it appears that the passage of some substances such as glycerol and urea may be facilitated. This has led to the concept that there might be aqueous pores present on a semi-permanent basis formed perhaps by proteins penetrating through the lipid and forming quasi-tubular connections from one side of the membrane to the other.

Evidence derived from the movement of water across red cell membranes has suggested that there could be pores 0.7 nm in diameter but the diffusion rates of various non-electrolytes indicate that if pores are present their diameter is likely to be smaller still, *c.* 0.56 nm. Furthermore, the electrical resistance of plasma membranes is usually in the range 1 000–10 000 ohms cm^{-2} or about one million times less than that of the body fluids. Consequently, if pores are present and if they permit the free passage of ions, we should not expect that their area should constitute more than 10^{-6} of the area of the membrane. However, since definite pores through the membrane cannot be seen in electron micrographs the question as to whether or not they actually exist must remain *sub judice* for the present. Probably we should not regard the plasma membrane as a rigid and stable structure at all. Time-lapse films of the frenetic activity of fibroblast cells growing in tissue culture provide a salutary lesson to anyone imagining a cell as a uniform structure surrounded by a permanent membrane. Measurements of the rate of turnover of the components of membranes (Table 6) reinforce the view that they exist in a dynamic rather than a static state. It would not be too facile to suppose therefore that at least some of the substances which penetrate into or out of cells do so during the replacement of membrane parts. Certainly it appears that the activity of secretory cells can be correlated with increased turnover of the constituents of the plasma membrane.

Table 6 Turnover times of different membrane lipids in myelin and mitochondrial membranes. (Adapted from O'BRIEN, J. S. (1967). *J. Theor. Biol.*, 15.)

	Average turnover time	
	Myelin	*Mitochondrial membranes*
Cerebroside	13 months	2 months
Sphingomyelin	10 months	1 month
Phosphatidyl choline	2 months	2 weeks
Phosphatidyl ethanolamine	7 months	4 weeks
Phosphatidyl serine	4 months	3 weeks
Phosphatidyl inositol	1.25 months	2 days

Discussion of the creation of pores by thermal movement and of the relatively rapid replacement of membrane components may leave the impression that membranes consist of a somewhat random association of individual molecules. Recent studies, far from supporting such a contention, suggest that within membranes there may be ordered aggregations of like molecules forming specific *domains* (Figs 4–3 and 4–4). Such aggregations or paracrystalline arrays may be of either protein or lipid. When composed of proteins, they have an obvious expression in the formation of the various types of cell junction. Lipid domains are also formed and G. L. Nicholson refers evocatively to the segregation of 'frozen' elements to form 'solid lipid icebergs floating in a fluid lipid sea' to describe the segregation which occurs when simple lipid mixtures are maintained at temperatures below the phase transition value of some of the components. The phase transition temperatures for many lipids are within the physiological range and consequently comparable processes may be involved in the sorting out of lipid domains in membranes.

Both individual molecules and domains can move laterally in membranes, though, as already mentioned, there is little relocation of lipids from one bilayer to the other.

Some proteins require relative freedom of movement in order to subserve their normal function. Thus, the enzyme Na–K–ATPase generally lies in membrane regions where the lipids are essentially mobile and unordered. Restrictions are, however, placed upon the free movement of some membrane components. Apart from the exclusion of non-homogeneous molecules from some lipid domains or the trapping of others within 'frozen' domains, constraints on membrane components, and in particular on the protein moiety, can involve

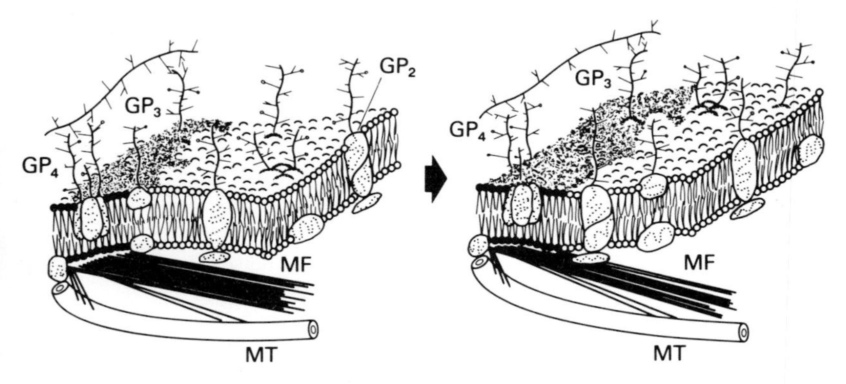

Fig. 4–3 Transmembrane control over the distribution and mobility of cell surface receptors by peripheral and membrane-associated cytoskeletal components. In this example the mobility of integral glycoprotein complexes GP_3 and GP_4 is controlled by outer surface peripheral components and also transmembrane linkages to membrane-associated cytoskeletal elements. In addition, complexes GP_3 and GP_4 could be sequestered into a specific lipid domain indicated by the shaded area, while complex GP_2 exists in a free or 'unanchored' state and is capable of lateral motion. MF, microfilaments; MT, microtubules. (From NICHOLSON, G. L. (1976). *Biochim. Biophys. Acta*, **457**.)

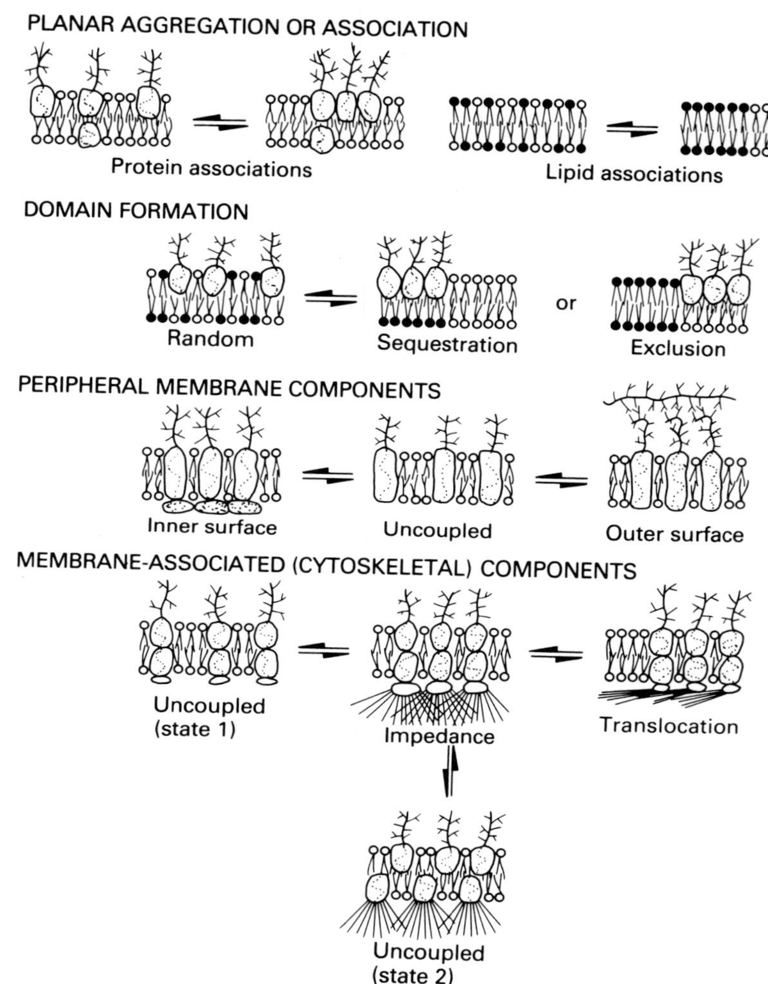

Fig. 4–4 Some possible restraining mechanisms on the lateral mobility of cell surface receptors. (From NICHOLSON, G. L. (1976). *Biochim. Biophys. Acta*, **457**.)

restrictions imposed by extra-membranal substances inside or outside the bilayer and by cytoskeletal structures (Fig. 4–3). In addition to the stabilizing effect produced by interlinking of the carbohydrate entities of glycoproteins peripheral to the plasma membrane, three types of structure, all on the cytoplasmic side of the membrane, influence the spatial distribution of membrane components. These cytoskeletal structures include (1) microfilaments (2) thick filaments and (3) microtubules. Bundles or sheets of microfilaments often lie in close approximation to the cytoplasmic face of plasma membranes whilst thick filaments and microtubules run in from the cytoplasm to interact, perhaps via Ca^{2+} as an intermediary, with proteins in the membrane, so that these molecules can be made to move relative to others. An example of such a process is seen in the 'capping' of lymphocytes. If the antigen receptors on the surface of lymphocytes are crosslinked with a fluorcein-labelled ligand they can be observed to relocate themselves on the cell surface, first into patches and then into a 'cap' at one pole. Once at the pole they are generally sloughed off and new receptor sites are incorporated as replacements over the membrane surface.

4.1.2 Cell recognition

All the cells of an individual apparently have certain characteristic proteins in their cell membranes which represent, as it were, a passport to acceptance within the community of cells within the body. In vertebrates, cells with the wrong recognition proteins are attacked by the antibody defence system. The small lymphocytes of the blood in mammals are largely responsible for the production of antibody proteins which attach to non-native cell proteins.

The antibody system provides a two-fold defence, firstly against foreign proteins or cells invading the body from outside and secondly against native cells which have mutated from their normal form. The value of the former is obvious in combating disease organisms; the latter, however, is a function of at least equal importance. One authority has estimated that as many as 10^6 mutations a day may occur in a human being. Many of these mutations are doubtless automatically lethal to the cells concerned but potentially the remainder could, if the mutated cells were permitted to survive, give rise to cell lines with functional malformations or malignant tendencies. However, only when a mutation does not result in a change in membrane protein can a mutated cell escape the attentions of the antibody system for long. Fortunately such concealed mutations are presumably rare or cancer would be more prevalent than it is. Significantly the incidence of malignancy tends to rise when the antibody system is immobilized.

It is, however, necessary to immobilize the antibody system after grafting foreign tissue or the host organism will reject it. Consequently, following transplant surgery a nice balance must be maintained to suppress the host's tendency to reject the foreign tissue without eliminating the ability to recognize and destroy disease organisms and malignant cells.

The plasma membranes of lymphocytes tend to be chemically differentiated in the sense that not all lymphocytes have the capability of responding to the presence of a given foreign protein in the blood. However, it is an important feature of the response to foreign molecules that lymphocytes which do interact with a particular molecule are also stimulated to divide. Processes occurring at the cell surface can therefore be implicated in the initiation of cell division.

Liver cells have four recognition (antigenic) proteins on their surface but little is yet known of the antigenic proteins of other tissues apart from blood cells. The antigens of blood cells, which constitute the familiar blood group factors, are mucopolysaccharides incorporating sialic acid and amino sugars. In the case of the Rhesus D factor there are about 3 000–20 000 sites of activity on the surface of each red cell. About 50 to 100 times as many sites for the A factor are present.

4.1.3 Surface interaction of cells

The cells of the blood system have a relatively independent existence but the majority of cells in multicellular organisms adhere to one another.

Adhesion of neighbouring cells is important in decreasing their motility. When cells such as fibroblasts are grown in tissue culture they move about by means of undulations passing along the surface of projecting pseudopodia. Cells meeting in the culture do not move over each other, instead they adhere at the point of contact and further movement in that region of the cells is suppressed. Some cancer (sarcoma) cells lack this contact inhibition and the ability to adhere to other cells. Consequently they are invasive and tend to penetrate between the cells of other tissues. In wound healing too the cells in the region of the damage lose some of their adhesiveness temporarily and move out of their normal position to assist in covering the damaged surface (Fig. 4–5). Changes in cell adhesion are important during embryogenesis and organ formation as cells move over one another into their correct location.

Adhesion between cells involves the establishment of intercellular junctions.

4.1.3.1 Interaction between plasma membranes. In tissues neighbouring plasma membranes are generally spaced some 20–30 nm apart but at specialized regions, the intercellular junctions, more direct contact is made between adjacent parts of the two membranes. Three main types of intercellular junction are recognized: (1) adhering junctions, (2) gap junctions and (3) tight junctions. The respective functions attributed to these three structures are mechanical adhesion, interchange of materials between cells and prevention of the free passage of materials from one side of an epithelium to the other via the intercellular channels.

4.1.3.2 Adhering junctions (Figs 4–6 and 4–1b). The principal mechanical interlinks between cells are formed by *desmosomes*, disc-like modifications of membranes usually some 0.2–0.5 μm in diameter. The adjacent plasma membranes remain about the usual distance apart, 22–35 nm, and the

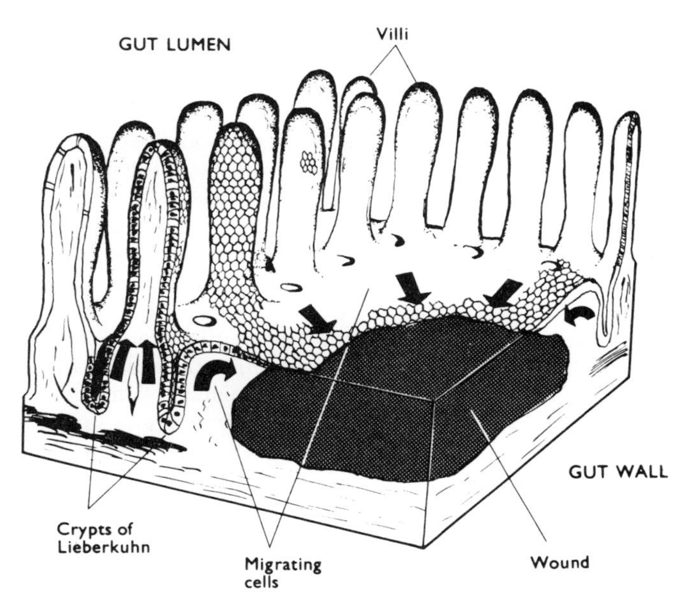

Fig. 4–5 Early stage of wound repair in the small intestine. Flattened epithelial cells derived in part from the crypts of Lieberkuhn are shown migrating out to cover the wound. In undamaged parts of the tissue cells originating in the crypts migrate instead along the sides of the villi to be lost eventually near the tips. (From JOHNSON, F. R. (1964). *Annual Reviews of the Scientific Basis of Medicine*. Athlone Press, London.)

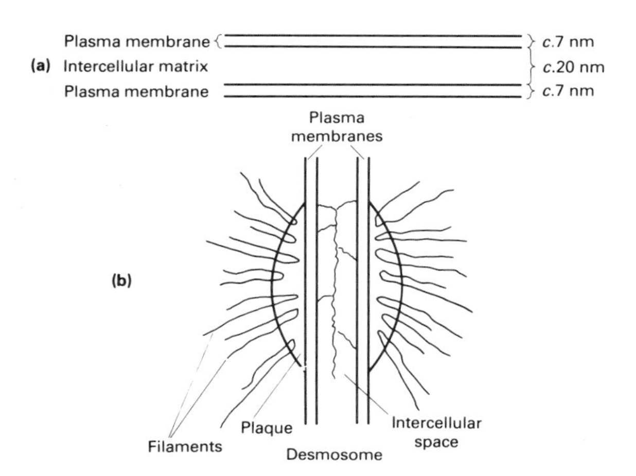

Fig. 4–6 (**a**) General dimensions of the plasma membranes and intercellular space. (**b**) Diagrammatic representation of a desmosome.

interspaces contain glycoproteins which, in effect, cement the membranes together. Sometimes there is an aggregation of this intercellular material in the mid-line between the two halves of a desmosome with additional cross links connecting the plasma membrane to this. On the cytoplasmic side of each plasma membrane in the desmosome region is a plaque which forms the point of attachment for fibrils (*c.* 10 nm diameter) which run in from the cytoplasm, loop round and return to the cytoplasm (Fig. 4–6b). The whole structure provides a means of attachment between neighbouring cells which is also capable of distributing mechanical stresses and strains through tissues.

Two types of desmosome have been described, the *macula adherens* in which the structure passively transmits mechanical forces between cells and the *fascia adherens* which, having contractile fibres composed of F–actin, can actively influence the shape of the neighbouring cell.

Desmosomes tend to be particularly frequent in tissues such as the skin which are exposed to mechanical stresses though they occur to some extent wherever cells interact. Separation of a tissue into its component cells tends to occur if desmosomes are destroyed by treatment with proteolytic enzymes and it may be noted in passing that cancerous cells tend to have low incidence of desmosomal links with neighbouring cells. Doubtless this factor contributes to their invasiveness.

4.1.3.3 Gap junctions (Fig. 4–7). Gap junctions, the site of interchange of inorganic ions and organic molecules of molecular weight up to about 10^3 daltons between neighbouring cells, occur at all levels of organization from the tissues of sponges and coelenterates upward through the animal kingdom. They are formed by adjacent plasma membranes coming together in close approximation so that the intercellular space is restricted to about 10% of its usual width. In addition, electron microscope studies suggest that groups of proteins forming hydrophilic channels through one plasma membrane come to be contiguous with comparable channels in the other. Studies of electrical resistance indicate that there is a much freer movement of ions between adjacent cells linked by gap junctions than across a cell membrane to the medium. Thus the resistance between neighbouring cells in *Drosophila* salivary gland is only of the order of 110–190 ohm cm^{-2} whereas that across a typical plasma membrane between cell contents and extracellular medium is in the range 500–10 000 ohm cm^{-2}. It is assumed that the current passing between adjacent cells in such experiments is carried by inorganic ions moving through the channels in the gap junctions.

The gap junctions serve as a means by which small molecules, inorganic ions, metabolites and even apparently nucleotides, can be distributed between neighbouring cells. In this way intercellular cooperation within a tissue may be improved by evening out the distribution of ions, amino acids and water whilst it also seems probable

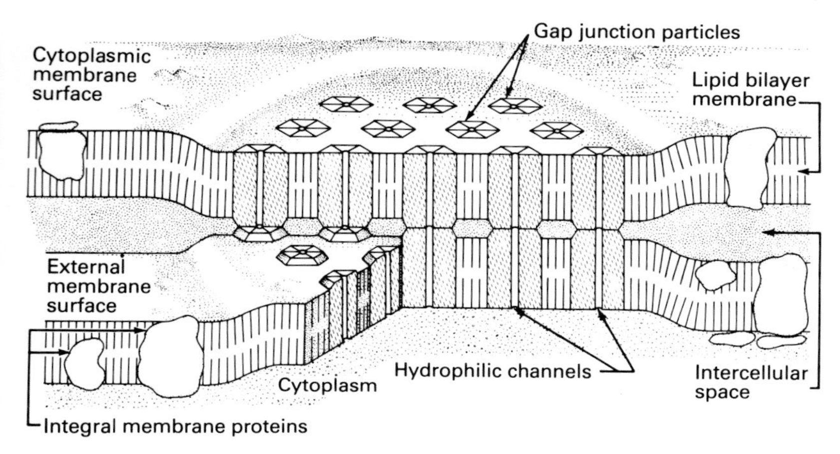

Fig. 4–7 Three-dimensional reconstruction of a gap junction. (From STAEHELIN, L. A. (1974). *Int. Rev. Cytol.*, 39.)

that sometimes genetic aberrations arising in some cell lines which should result in metabolic pathway blocks may not have the deleterious effects to be expected because of the ability of the cells to acquire the necessary post-block substrates from non-mutant neighbouring cells.

A good example of intercellular cooperation via gap junctions is provided by embryonic heart cells. When cultured in isolation and then allowed to come into contact with one another they begin to contract synchronously within half an hour. This timing agrees well with the period necessary to establish gap junctions. However, if heart cells are linked only via L-cells (which do not form gap junctions) no synchronous beat is achieved.

4.1.3.4 Tight junctions (Figs 4–1a and 4–8). Tight junctions, found particularly near the apical region of epithelial cells, form partial barriers to the free passage of water and inorganic ions along intercellular channels.

Typically the tight junction is formed by a series of line of direct contact between adjacent plasma membranes. Along each such contact strand common elements so link the two membranes that the overall thickness is less than the sum of the thickness of the two membranes in unlinked regions. Unlike the gap junctions and desmosomes, which are relatively discrete points of contact, the tight junctions extend across the whole interface between one epithelial cell and its neighbour thus closing off free access along the intercellular channel between one side of the epithelium and the other (Fig. 4–9). Tight junctions probably completely prevent the passage of large molecules such as proteins. They are not generally thought to be impervious to inorganic ions and water but

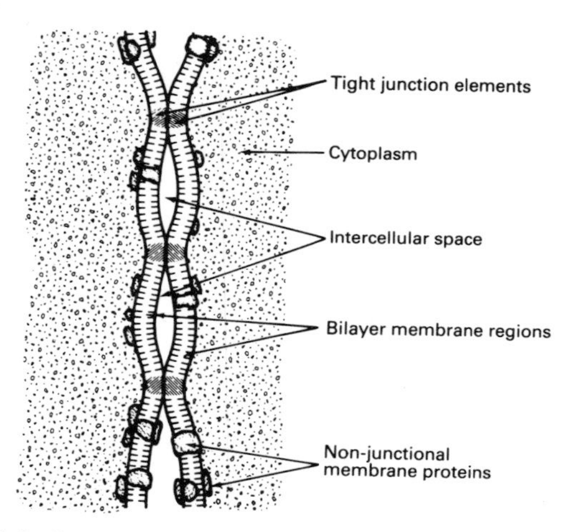

Fig. 4–8 Model of a tight junction. The simplified bilayer membranes are held together along lines of attachment by tight junction elements (proteinaceous particles), which are depicted within the membranes as striated areas. Each of these elements seems to bridge the width of the membrane in which it resides, and appears attached to its counterpart in the plane of the intercellular space. (From STAEHELIN, L. A., (1974). *Int. Rev. Cytol.*, **39**.)

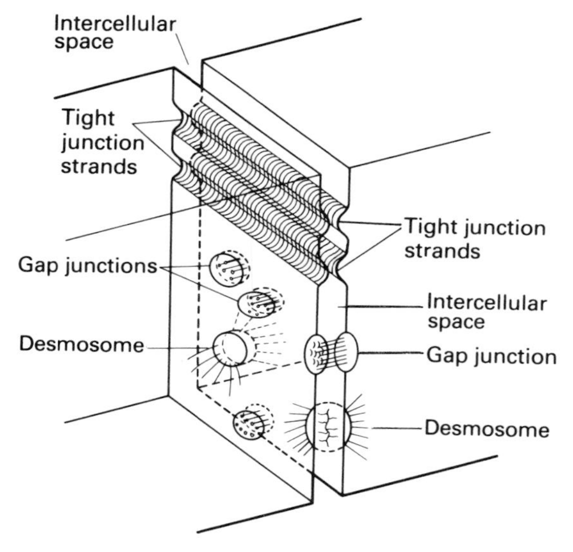

Fig. 4–9 Diagram to illustrate that whilst gap junctions and desmosomes form limited points of contact between cells, tight junctions form a continuous barrier across the intercellular space. N.B. The various forms of junction are not to scale being drawn larger than their true size relative to the cell.

nevertheless are sufficiently impermeable to play an important passive role in the maintenance of concentration gradients across epithelia created initially by ion transport systems in the plasma membranes (cf. Chapter 6). Correlation between structure and function occurs in the sense that the less leaky a tight junction, the greater are the number of crosslinking strands between the cells. Relatively non-leaky epithelia, such as that of the urinary bladder of the toad, have about eight strands whereas more pervious tissues, e.g. the proximal tubule of *Necturus* nephrons, have only about three strands per junction.

4.2 Endoplasmic reticulum

The endoplasmic reticulum (ergastoplasm) is characteristically present in all cells except mature anucleate red blood cells but it is particularly well developed in cells which secrete proteins.

Two types of reticulum are recognized:

(1) Rough endoplasmic reticulum (rough-surfaced reticulum, granular reticulum, RER) (Figs 4–1, 4–10, 4–11).
(2) Smooth endoplasmic reticulum (smooth-surfaced reticulum, agranular reticulum, SER) (Fig. 4–11).

Rough endoplasmic reticulum is composed of a system of interconnecting tubules some 40–70 nm in diameter and usually these are closely applied to one another in certain regions so that the tubules are flattened and appear in sections as parallel arrays of membranes. The lumena of the tubules in these regions are termed *cisternae*.

Rough endoplasmic reticulum is so called because the tubules bear particles, ribosomes, on the outer surface. The ribosomes, each some 15 nm in diameter and shaped like a cottage loaf, consist of ribonuclear protein. In combination with messenger RNA the ribosomes form the essential machinery for protein synthesis for it is at these sites that amino acids, linked to specific transfer RNAs, are assembled into proteins. After assembly the proteins, or some of them, pass across the reticular membrane into the tubule lumen. As would be expected from its role in protein formation the rough endoplasmic reticulum proliferates in the cells of tissues, such as the pancreas and spider silk gland, which produce protein on a large scale.

Smooth endoplasmic reticulum differs from the former type in (a) lacking ribosomes and (b) rarely forming cisternae. This type of reticulum hypertrophies in cells, such as those of the testis and adrenal cortex, which are involved in the synthesis of steroids. The SER also appears to function in cholesterol metabolism and perhaps in lipid metabolism in general. Connections are sometimes apparent between smooth and rough reticula and it seems that the former is implicated in the transfer of some of the enzymes fabricated in the ribosomes from the cisternae of the reticulum to the Golgi apparatus. This transfer appears to be associated with a flow

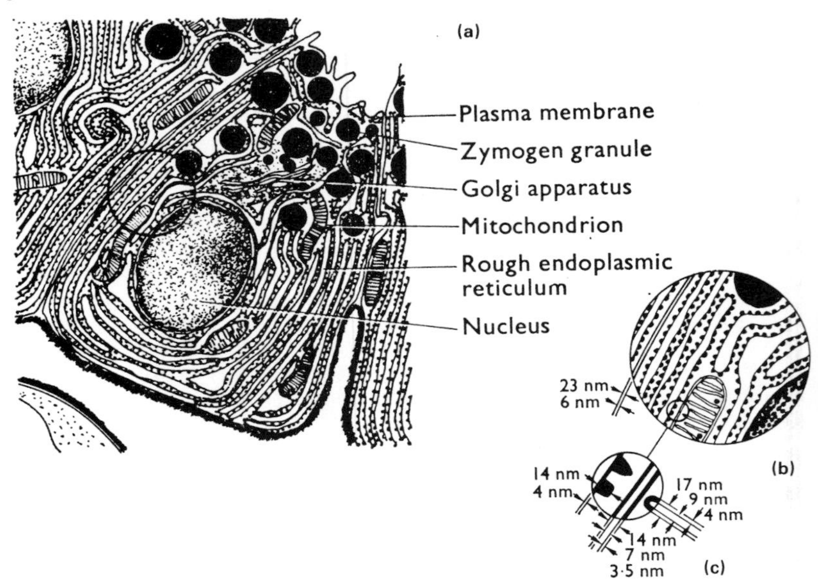

Fig. 4–10 Diagram to illustrate the dimensions of various cellular membranes in the secretory cells of mouse pancreas. (**a**) General topography of the cell. (**b**) Dimensions of plasma membrane and combined plasma membranes plus intercellular matrix. (**c**) Dimensions of rough endoplasmic reticulum membrane, ribosomes, outer and inner boundary membranes of a mitochondrion and of the cristae. (From SJÖSTRAND, F. J. (1956). *Int. Rev. Cytol.*, 5.)

of membrane from one system to another. Direct budding of vesicles from the rough endoplasmic reticulum also occurs as an alternative to transfer via the smooth reticulum.

4·3 Nuclear envelope (nuclear membrane) (Fig. 4–11)

Except at the time of cell division the nucleus of a cell is separated from the cytoplasm by the *nuclear envelope*. This is composed of parallel membranes, each 7–8 nm thick enclosing a space 30–50 nm across. The paired membranes join in places to form octagonal pores some 65 nm in diameter (Fig. 4–12). The total number of pores varies with the size of the nucleus (in yeast cells there are about 2000) but the dimensions and form are relatively similar in such unrelated species as newt, starfish and frog.

The outer surface of the nuclear envelope frequently bears ribosomes similar to those on the endoplasmic reticulum and it has been postulated that the nuclear membrane may represent just a specialized region of the reticulum. This view receives additional support from electron micrographs which seem to indicate that when the nuclear envelope

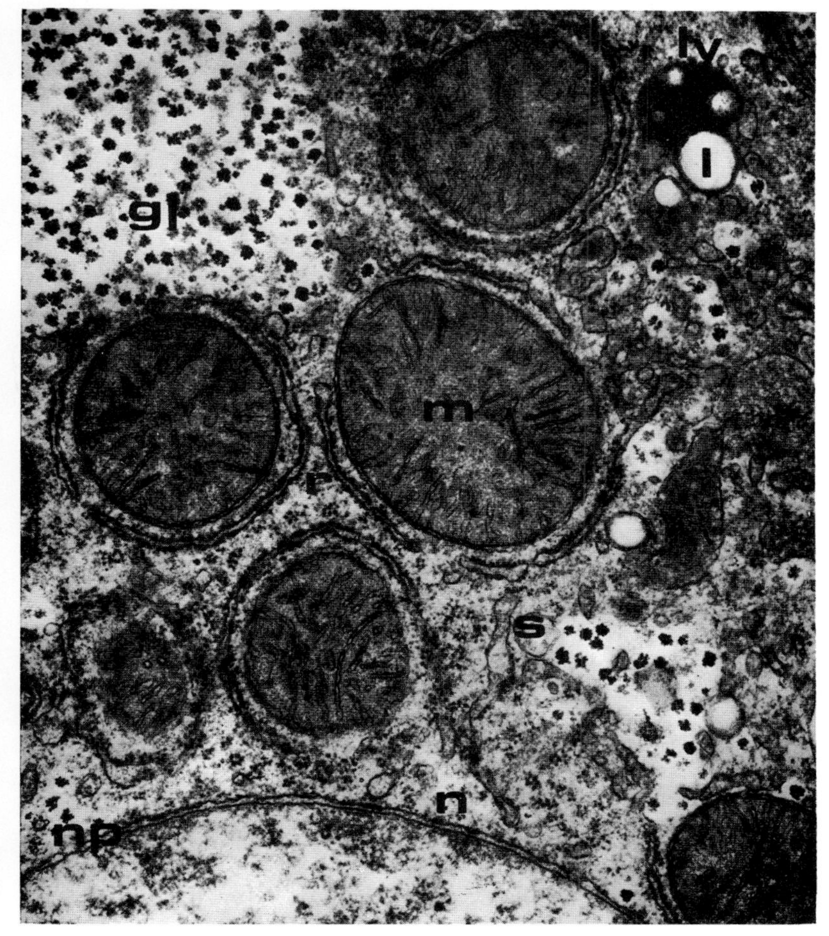

Fig. 4–11 Portion of bat liver cell, gl, glycogen granules; l, lipid; ly, lysosome; m, mitochondrion; n, nuclear membrane; np, nuclear pore; r, rough endoplasmic reticulum; s, smooth endoplasmic reticulum. (× 40 500.) (Reproduced by courtesy of Mr R. P. Gould.)

disintegrates at the time of cell division it forms vesicles similar to those of the reticulum. Furthermore, the nuclear envelope is reformed after nuclear division from the components of the rough-surfaced reticulum.

4.4 Annulate lamellae (Fig. 4–13)

This membranous system, which resembles the nuclear membrane in having pores arranged in hexagonal array, occurs in the cytoplasm of

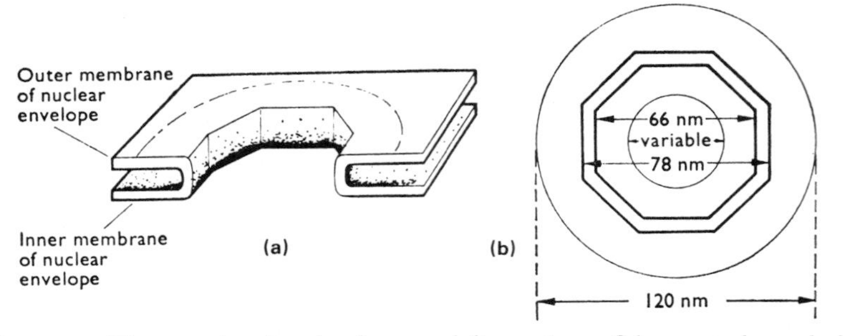

Fig. 4–12 Diagram showing the shape and dimensions of the pores through the nuclear envelope. (Slightly modified from GALL, J. G. (1967). *J. Cell Biol.*, 32.)

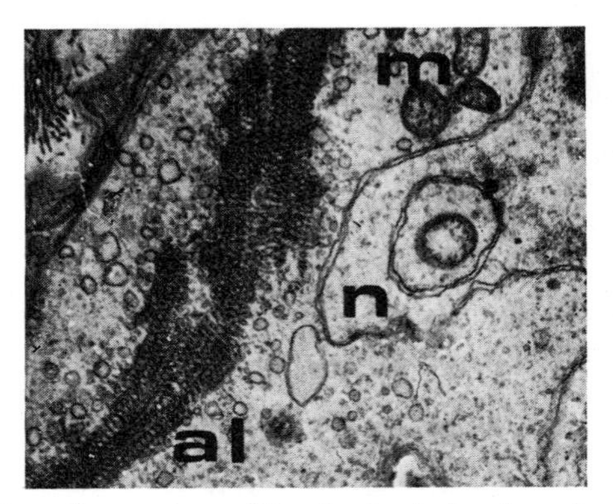

Fig. 4–13 Portion of oogonium of *Xenopus laevis* showing, al, annulate lamellae; n, nuclear membrane. (Reproduced by courtesy of Mr A. C. Webb.)

oocytes and some tumour cells but is rarely found elsewhere. Annulate lamellae are formed by the fusion and reorganization of 'blebs' budded off from the nuclear membrane. Little is known of their function.

4·5 Golgi apparatus (Fig. 2–2)

The Golgi apparatus (Golgi system, dictyosomes, y-cytomembranes) consists of a filamentous or plate-like reticulum of smooth membranes (Fig. 2–2). The organelle appears to be present in all cells and reaches its maximum size and complexity in secretory and storage tissues. In the pancreas of guinea-pigs, for example, the Golgi apparatus occupies as much as 6–10% of the cell volume and has three regions, cisternae, central vacuolar region and peripheral vesicles. A comparable degree of complexity is found in the corresponding structure in plant cells (Fig. 4–14). The model figured shows five flattened cisternae from which anastamozing tubules extend peripherally. Two types of vesicle are in

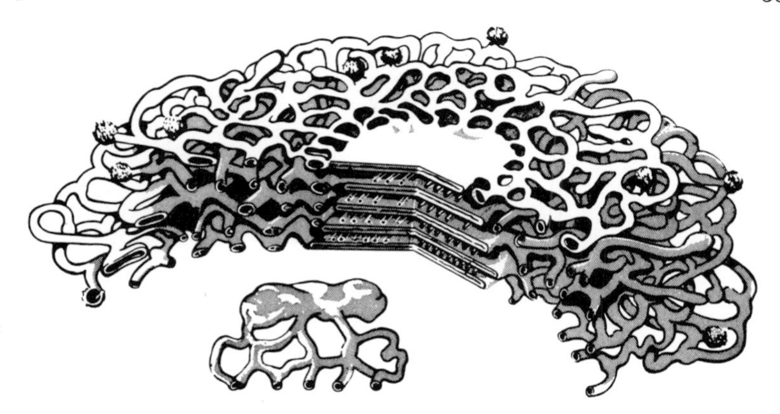

Fig. 4–14 Diagrammatic representation of a plant dictyosome (Golgi apparatus) made up of five cisternae. Note the microtubules between the cisternae. Insert shows the formation of a secretion vesicle from the Golgi. (From MOLLENHAUER, H. H. and MORRÉ, D. J. (1966). *Ann. Rev. Plant Physiol.*, **17**.)

the process of formation from the tubules and arrays of microtubules lie between the cisternae.

Golgi membranes are smooth, some 7 nm thick and show the unit membrane form in electron micrographs. Successive membranes forming the walls of the cisternae are about 14 nm apart, closer than in the cisternae of the endoplasmic reticulum.

Characteristic enzymes of Golgi membranes are those linking sugars with protein to form glycoprotein. For example, the mucopolysaccharide used by *Hydra* for attaching its base to the substratum originates in Golgi vesicles. However, in addition to acting as a surface for enzyme binding the Golgi membranes are also used to package enzymes which are to be transported about the cell. The encapsulating of both primary lysosome containing enzymes for intracellular use, and granules containing the precursors of extracellular digestive enzymes occurs here.

4.6 Lysosome system (Figs 4–11 and 4–15)

4.6.1 The components of the lysosome system

Autolysis of parts of the cell which are no longer required and digestion of macromolecules taken in across the cell surface are the primary responsibility of the system of membrane-bounded vesicles known as lysosomes. There are three principal components in this intracellular digestive complex, *primary lysosomes* which contain digestive enzymes; *heterophagosomes* enclosing macromolecules or particulate material taken in across the cell surface and *autophagosomes* enclosing cell organelles due for destruction.

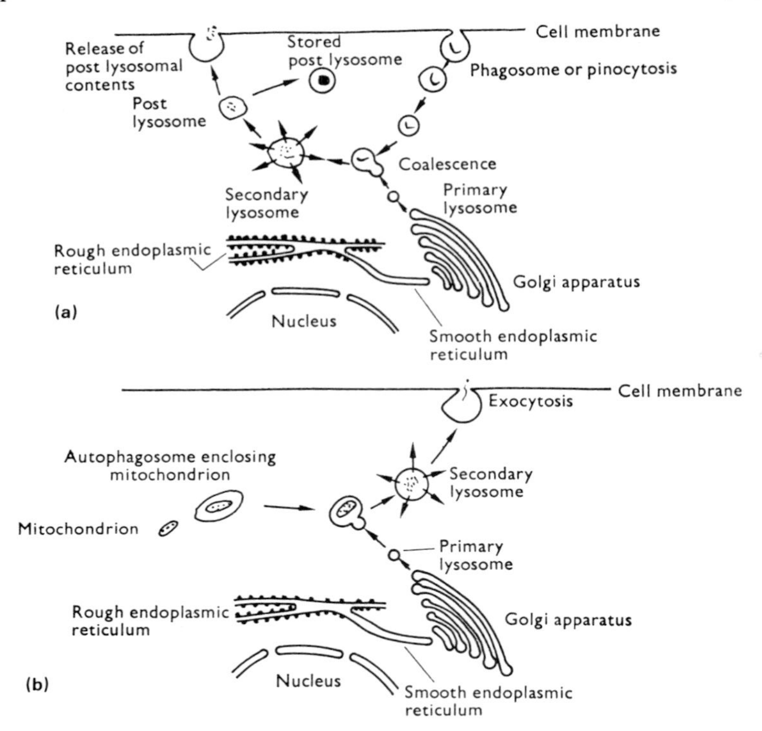

Fig. 4–15 Diagram illustrating the probable mode of formation and subsequent fate of lysosomes. (**a**) Sequence of (1) formation of primary lysosomes, (2) coalescence with vesicles formed by pinocytosis or phagocytosis to give secondary lysosomes, (3) resorption of useful material into the cytoplasm, and (4) storage or release of post lysosomes. (**b**) Formation of autophagosome vesicles and their subsequent digestion by enzymes from primary lysosomes. (Based on descriptions by ALLISON (1968) and DE DUVE and WATTIAUX (1966).)

When primary lysosomes are brought into contact with either heterophagosomes or autophagosomes the membranes of the two vesicles coalesce bringing the digestive enzymes in contact with the substrate material. The compound vesicle so formed is termed a *secondary lysosome*. As digestion proceeds within this latter vesicle products capable of being handled by the cytoplasm escape across the bounding membrane. Digestion finally ceases leaving a *post lysosome* which may either remain within the cell or alternatively release its contents through the plasma membrane.

Primary lysosomes are in effect pre-packaged bags of enzymes with the capacity to break down a wide range of substances including proteins, ATP, DNA and RNA. Altogether about 15 enzymes have been recognized in lysosomes so far including acid phosphatase, cathepsins (proteolytic enzymes), acid RNAase, acid DNAase and α-glucosidase. Since these enzymes will break down components of the general cytoplasm and nucleoplasm just as readily as they will substrates contained within the heterophagosomes and autophagosomes it is of course essential for the well-being of the cell that they be isolated in impervious membrane-bounded vesicles. They are in fact surrounded by a unit membrane some 9 nm across, somewhat thicker than that of most cell organelles.

Primary lysosomes are thought to arise by the budding off of vesicles from the Golgi apparatus though probably the enzymes are elaborated initially in the ribosomes attached to the endoplasmic reticulum.

Heterophagosomes. Protozoa, such as *Amoeba*, and phagocytic cells of the blood and reticulo-endothelial systems in higher organisms can surround, ingest and subsequently digest particulate matter. 'Food' vacuoles (heterophagosomes) formed by such ingestion of particles may be 10 μm or more in diameter.

Smaller vacuoles containing ingested medium are produced in blood cells and *Amoeba* by a process termed *pinocytosis* (cell drinking). Pinocytosis is not continuous in *Amoeba* but will begin if the organism is placed into a dilute (1.5%) solution of protein. The cell membrane then intucks in one or more places to form narrow channels. At the lower end of these channels small vesicles (*c.* 1 μm in diameter) separate off and are carried into the cytoplasm. In higher organisms cells with brush borders form pinocytotic vesicles at the base of the 'brush'.

In some other tissue, though not in muscle or nerve, still smaller (micropinocytotic) vesicles only 0.1 μm in diameter, are formed as depressions in the cell membrane rather than at the bottom of channels.

Although of different sizes the micropinocytotic vesicles, pinocytotic vesicles and phagosomes all have in common the fact that the membrane which surrounds them is derived from the plasma membrane and that the liquid which they contain initially is extracellular medium. Hence they may all be grouped under the common heading of heterophagosomes.

Pinocytosis involves more than just mere enclosure of a droplet of the extracellular medium. A variety of substances, including sucrose, non-ionic detergents, foreign protein, mannitol, etc., can all be concentrated in the vesicles to a level higher than that in the extracellular medium. It is presumed that this concentration is due to an adsorption of the substance concerned onto the cell membrane prior to its invagination to form the pinocytotic vesicle.

Autophagosomes. H. Grundfest's remark 'In its basic outlines the economy of living cells is marvellously parsimonious, old junk is modified and fashioned to new purposes', although coined for another reason, aptly describes the function of autophagosomes since the role of these structures is the degradation of (surplus?) organelles so that their components may be re-utilized. Autophagosomes are formed by the segregation of a single organelle or a region of the cytoplasm by a smooth membrane.

When either an autophagosome or a heterophagosome comes into contact with a primary lysosome the membranes of the two vesicles coalesce and a secondary lysosome is formed.

Secondary lysosomes. Digestive enzymes from the primary lysosomes break down the material brought into the cells by the heterophagosomes or the contents of autophagosomes and the small molecules resulting from these lytic processes escape across the bounding membrane into the general cytoplasm. When digestion is complete the vesicle and its remaining contents is termed a *post lysosome*.

Post lysosomes have two possible fates. Either they remain as relatively inert bodies for long periods or alternatively their membranes may link with the cell membrane and release the undigested contents to the extracellular medium. Which of these alternatives is followed depends on the nature of the host cell and on the opportunities for the removal of the products of such cell 'defaecation'. Liver cells have a ready means of removing lysosomal waste by depositing it in the bile and indeed there is a tendency for post lysosomes in liver cells to be concentrated near the bilary canals. By contrast with liver cells, skin cells tend to retain post lysosomes, a feature which is made use of in tattooing since carbon particles taken into the lysosomes remain in the cell for long periods.

4.6.2 Specialization of the lysosomal system

The intracellular digestive system described above involves, in essence, the extrusion of enzymes (from the primary lysosome) into an isolated part of the extracellular medium (lumen of the heterophagocyte). Clearly, by a slight alteration of the timing of the process the primary lysosomes could be made to discharge into pinocytotic vesicles before they separate from the cell surface or even directly through the surface itself and so exert an influence on the tissues surrounding the cell rather than on materials taken in. It is easy to draw an analogy between this theoretical modification of the lysosomal system and the observed extracellular secretion of enzymes by some cells. For example, the zymogen granules of the pancreatic secretion arise from the Golgi apparatus and are carried to the cell surface in membrane bounded vesicles prior to release (Fig. 4–10).

Similarly the extracellular enzymes of osteoclasts, the cells responsible for breaking down unwanted bone tissue, also pass to the cell surface in lysosome-like vesicles.

Pinocytosis is also open to modification of function. Potentially a heterophagosome is capable of transporting material in at one face of a cell and out at another, i.e. going in effect from the process of pinocytosis to excretion as a post lysosome without having received digestive enzymes from a primary lysosome. Some such process appears to be involved in the transport of serum from the blood across the endothelial cells lining capillaries to the interstitial space and also in the transference of antibody proteins from the maternal blood to that of the foetus across the wall of the placenta.

Modification of autophagosome activity is principally associated with metamorphosis (particularly in insects and amphibia) when extensive cytoplasmic degradation occurs.

4.6.3 Lysosomes and disease

Speaking of the potential hazard to the cell of the lysosomal enzymes Christian de Duve has referred to them as being 'safely behind bars' in the lysomal vesicles. On the death of a cell these membrane 'bars' are broken down and the lysosomal enzymes escape into the cytoplasm. Rapid and irreversible autolysis of the cell contents follows. During life therefore the survival of a cell depends on its ability to maintain the integrity of the lysosomal membranes.

Various substances, including silica, carbon tetrachloride, croton oil and polyene antibiotics are known to increase the fragility of lysosomal membranes and are therefore potentially capable of causing the destruction of any cell into whose heterophagosomes they are taken. Thus silica particles taken up by macrophage cells can destroy them. Subsequently the silica is released and if taken up by other macrophages will result in their death in turn. In miners, uptake of silica particles into the lungs is often extensive and the subsequent destruction of many macrophages coupled with the fact that collagen tends to be laid down around the dead cells, can lead to the impairment of lung function and the consequent onset of the condition known as silicosis.

Some of the polycyclic hydrocarbons and dyes which are known to be able to give rise to cancerous conditions also have the ability to disrupt lysosomal membranes and the possibility must be considered that they owe their carcinogenic activity to the release of DNAse from lysosomes and that this enzyme secondarily brings about mutations which give rise to the cancerous state.

4·7 Microbodies

Microbodies (peroxisomes, uricosomes, cytoplasmic bodies type 1) are discrete, usually spherical or ovoid bodies, bounded by a single membrane 6.5 nm thick. They are known to occur in the cells of a wide range of organisms including plants, protozoa and multicellular animals. In higher organisms they are best known from kidney and liver cells. The principal enzymes associated with these organelles are catalase, urate oxidase and D-amino acid oxidase though only the first of these is invariably present.

Microbodies usually have a diameter of approximately 0.3–0.6 μm and, in the case of the liver, some 400–800 are present in each cell. They appear to be derived from the endoplasmic reticulum or possibly the Golgi apparatus but should not be confused with microsomes which are artefacts of generally similar size and membrane appearance created by disruption of the endoplasmic reticulum during centrifugation.

Microbodies are concerned in amino acid and uric acid metabolism but the precise metabolic role of their catalase enzyme remains uncertain.

4.8 Microtubules (Figs 4–16, 4–17 and 4–18)

Thin, cylindrical membrane systems (microtubules) of approximately similar dimensions occur in many cells but the wide variety of their positioning and apparent function suggest that it is unlikely that they form parts of a common system. Examples of their occurrence include: the fibrillar infrastructure of cilia, flagellae and centrioles; the fibrillar elements of the spindle and asters of the mitotic figure at the time of cell division, a sub-surface band round the equatorial region of red blood cells (newt); lying close to the cleavage furrow in dividing cells; underlying the pellicular structure in *Paramecium*; lying free and not orientated in any apparent fashion in the cytoplasm of *Amoeba*.

Typically, microtubules are about 20–25 nm in diameter with a membranous outside wall 5–10 nm thick and an internal lumen 10–20

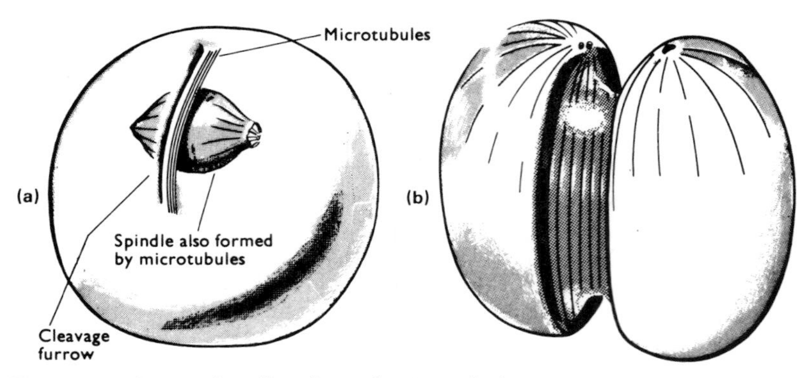

Fig. 4–16 Supporting function of microtubules. (**a**) Diagram showing the disposition of a band of microtubules in relation to the cleavage furrow at metaphase during cell division in *Chlamydomonas*. The mitotic spindle is also shown. (**b**) Microtubules radiating from the region of the basal bodies and orientated with respect to the cleavage furrow at a later stage in the division of *Chlamydomonas*. ((**a**) and (**b**) from JOHNSON, U. G. and PORTER, K. (1968). *J. Cell Biol.*, 38.)

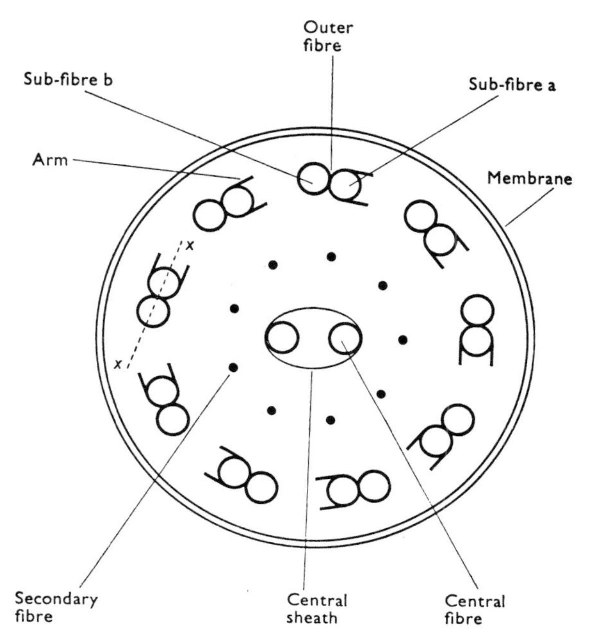

Fig. 4–17 Diagram of the '9 + 2' system of microtubules in a flagellum from the flagellate protozoan *Trichonympha*. x –––––– x is the axial plane of a microtubule doublet. (From GIBBONS, I. R. and GRIMSTONE, A. V. (1960). *J. Biophys. Biochem. Cytol.*, **7**.)

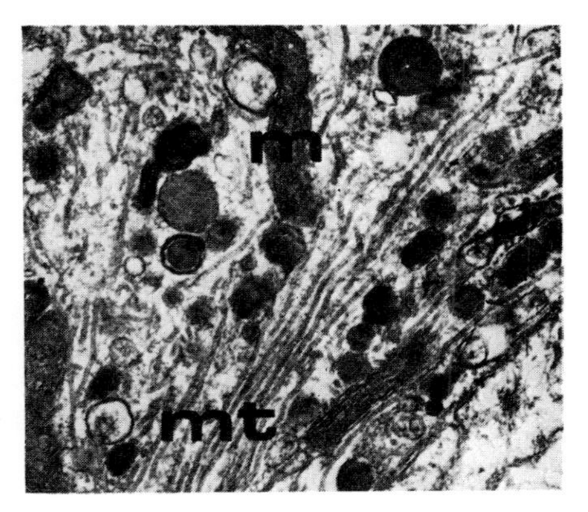

Fig. 4–18 Portion of nerve terminal in rat posterior pituitary showing microtubules (neurotubules), m, mitochondrion; mt, microtubules. (× 28 000.) (Reproduced by courtesy of Mr R. P. Gould.)

nm across. Situated longitudinally in the wall are a number of sub-fibrils (12–13) each about 4 nm in diameter.

The variety of their positioning suggest that the functions of microtubules may include a guiding of cytoplasmic streaming movements and providing structural support. It has also been suggested that the microtubules of cilia may slide upon one another to produce the ciliary beat. Microtubules in chromatophores are thought to be involved in the movement of pigments during colour change.

The endoplasmic reticulum is probably responsible for the initial assembly of microtubules.

4.9 Mitochondria (Figs 4–11 and 4–19)

These organelles are often referred to as the 'power houses' of cells because they are responsible for the production of most of the high energy donor substance adenosine triphosphate (ATP). Individual mitochondria, though most commonly ellipsoid, occur in a wide variety of shapes. In size they are commonly 1–10 μm long by 0.5 μm in width. The number per cell ranges from 50 to several thousand depending upon cell size and rate of energy expenditure.

Although so small, these organelles have quite a complex internal structure. The major units of this consist of a double boundary membrane to the inner wall of which are attached a series of membranes known as cristae (Fig. 4–19). Each of these has been described as resembling an empty rubber hot-water bottle, the neck of which is attached to the inner boundary membrane so that the lumen is in open connection with the space between the inner and outer walls of the boundary membrane (Figs. 4–19b and c). The number of cristae per mitochondrion varies according to the metabolic activity of the tissue from which the organelle comes. Mitochondria from heart cells, which have high respiratory rates, contain more than do those from the metabolically less active cells of the liver.

After osmium fixation and negative staining the cristae, but not the outer boundary membrane, bear arrays of projecting structures consisting of a stalk and headpiece mounted on a basepiece which is itself part of the cristal membrane (Fig. 4–19d and e). Together the basepiece, stalk and headpiece are termed a *repeating unit*. In beef heart mitochondria there are about 2000–4000 units μm^{-2} of cristal membrane and a total of some 10^4–10^5 in the whole mitochondrion. It is still uncertain as to whether or not the headpiece and stalk actually project from the membrane in the living organism. Since the headpiece is not separate from the membrane in freeze-etched preparations it may be that the osmium-fixed preparations show a pop-up artefact untypical of the living membrane. Nevertheless it is possible to separate the components

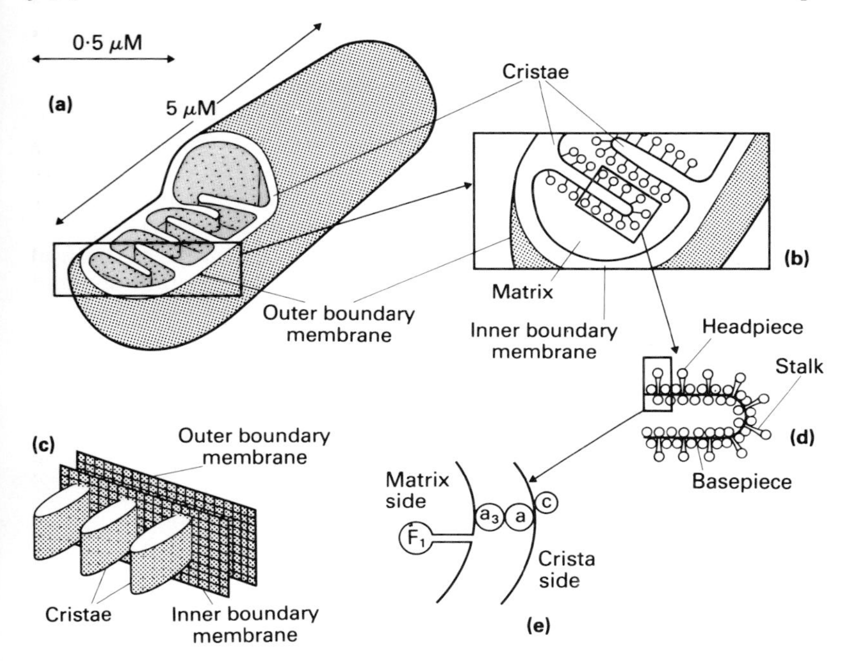

Fig. 4–19 The fine structure of a mitochondrion. (**a**) Model of a mitochondrion cut away to show the cristae inside. (**b**) Enlargement of a section of (**a**) to illustrate the cristae in more detail. (**c**) Relationship between cristal membranes and the outer and inner boundary membranes. Note headpieces covering the surface of the cristae. (**d**) Enlargement of crista showing the headpieces and basepieces of which the membrane is composed. (**e**) Location of enzyme systems in the basepieces (F_1 – coupling protein involved in ATP production). ((**c**) from PENNISTON, J. T. *et al.* (1968). *Proc. Nat. Acad. Sci.,* 59; (**e**) slightly modified from RACHER, E. (1974). In *Dynamics of Energy Transducing Membranes.* Eds. L. Ernster, R. W. Esterbrook and E. C. Slater. Elsevier Scientific, Amsterdam.)

of repeating units from living membranes and assign functions to the various parts.

Each mitochondrion contains at least 50 different types of enzyme and most of these are involved in the complex series of degradations responsible for the controlled transfer of bond energy from metabolic substrates to ATP. The primary metabolic substrate for energy production is glucose though the products of fatty acid and amino acid metabolism can both be utilized by mitochondria. The degradation and oxidation of glucose to CO_2 and water with production of ATP takes place in a large number of stages but these may be grouped into three main processes, glycolysis, the Krebs cycle (oxidative decarboxylation cycle) and oxidative phosphorylation. Of these only glycolysis occurs outside the mitochondrion.

This is not the place to go into the biochemical details of the processes concerned in oxidative decarboxylation and oxidative phosphorylation, though it can be seen from the diagram (Fig. 4–20) that they are complex. It is the function of the membrane systems in mitochondria to provide the necessary supporting structure so that the enzymes responsible for successive steps in the process can be grouped in such a manner that the reaction proceeds with the maximum of expedition. Close association of the various elements comprising the cytochrome chain permits the rapid transfer of electrons from one to the next. (Figs 4–19 and 4–20.)

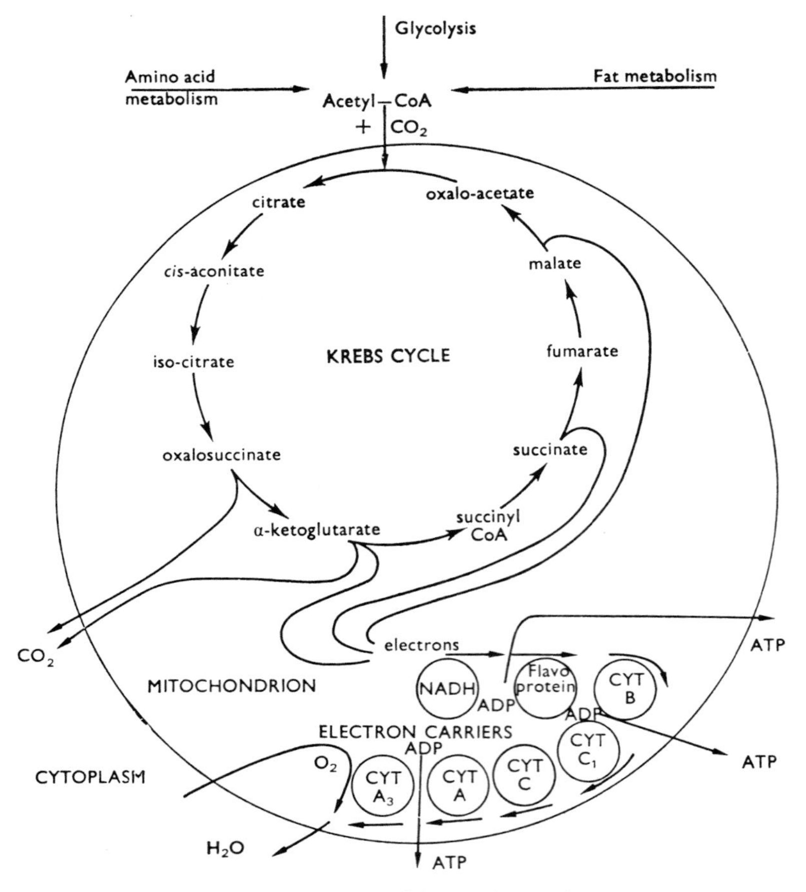

Fig. 4–20 Diagram to illustrate some of the biochemical processes occurring in mitochondria.

The functions of the various parts of mitochondria have been studied after separation of the different regions by techniques involving selectively breaking the membranes with emulsifying agents such as bile salts, detergents and distilbesterol or ultrasonics. The resultant membrane fragments are then fractionated by ultra-centrifugation for biochemical analysis.

By such means it has been shown that about 30% of the total protein is in the boundary membrane and 70% in the cristae. About half the protein is relatively constant in composition and molecular size irrespective of the source of the mitochondrion and this may represent structural protein. The remaining, enzymic, fraction of the protein is divided up between the various regions.

Controversy still surrounds the location of the dehydrogenase enzymes involved in the Krebs cycle though present indications suggest that they are located on the inside of the inner boundary membrane and that separate carrier mechanisms are involved in the transference of metabolites across the boundary membrane. The inner boundary and cristal membranes also bear the enzymes associated with oxidative phosphorylation. A coupling enzyme required for the production of ATP is located in the headpieces and the cytochromes of the electron transfer chain are in the basepieces of the repeating units. The precise arrangement of the enzyme complexes involved in electron transfer is still open to question but some recent evidence suggests that basepieces are subdivided into two layers; the outer layer is believed to contain cytochromes a and a_3 and the inner layer incorporates structural protein together with cytochromes c and c_1 (Fig. 4–19e).

Mitochondria are often found in close association with elements of the rough endoplasmic reticulum and it is possible that at least some of the components of the outer boundary membrane are produced by this organelle.

Mitochondria have a measure of autonomy within the cell and possess their own DNA. Mitochondrial DNA is, however, unlike that of the nucleus in that the molecules are ring-shaped and attached to the mitochondrial boundary membrane. This condition, which resembles that found in some bacteria, has led to renewed interest in the entertaining hypothesis put forward in the last century that mitochondria may have originated as independent organisms which then came to have a symbiotic relationship with primaeval cells and have subsequently extended the relationship until both cell and mitochondrion are mutually dependent.

5 Origin, Relationships and Turnover of Membranes

Membrane synthesis may be considered to take place in three phases: (1) the origin of the component proteins and lipids, (2) the assembly of the lipids and protein into a membrane and (3) constitution of the formed membrane elements into their functional position and state.

Proteins are synthesized in ribosomes and fatty acids are formed (or at least fatty acid chains are extended) by enzymes on the envelope of mitochondria. Subsequently incorporation of fatty acids into phospholipids occurs in the microsome fraction of homogenized cells. Since microsomes are largely derived from endoplasmic reticulum this membrane system is presumed to be primarily responsible for the production of phospholipid, though studies on the incorporation of ^{32}P suggest that phospholipid can also be made in the Golgi apparatus. The rate of formation of phospholipid seems to be related to the amount of membrane protein available and production ceases if insufficient protein is present.

Assembly of preformed phospholipids and proteins to form rough endoplasmic reticulum occurs in the existing reticulum and smooth endoplasmic reticulum is probably formed from the rough reticulum either by loss of ribosomes from the latter or directly as a smooth membrane synthesized between the ribosomes.

Smooth endoplasmic reticulum often extends close to the Golgi apparatus and vesicles budded from the reticulum may fuse with the Golgi vesicles but it is unlikely that the membranes of the two organelles as a whole are strictly homologous or that membranous vesicles derived from the reticulum are the sole source of Golgi membranes.

If fusion of membranes can be taken as an indication of close relationship, then it appears that the cell membrane has more in common with the Golgi membranes than it has with the endoplasmic reticulum. Though there is some argument on the point, it now seems unlikely that the endoplasmic reticulum ever joins with the plasma membrane. By contrast Golgi vesicles give rise to the membranes surrounding primary lysosomes and these can fuse with phagosomes or pinocytotic vesicles (whose membranes are derived from the plasma membrane).

The endoplasmic reticulum seemingly bears a close relationship to nuclear membrane. The latter is known to bud off vesicles which subsequently reorganize to form annulate lamellae in egg cells. These annulate membranes are occasionally found to be continuous with the rough endoplasmic reticulum. Furthermore, not only does the nuclear membrane often bear ribosomes but when it breaks up during mitosis it gives rise to membranous vesicles resembling endoplasmic reticulum. Conversely the new nuclear membranes formed after completion of cell division are derived from endoplasmic reticulum.

The inner membranes of mitochondria and chloroplasts stand distinct from the other cytomembranes of the cell. They are formed by incorporation of new material into existing membranes followed by self-replication of the organelle. If all of either type of organelle are lost from a cell new ones cannot be synthesized *de novo*; centrioles and basal bodies are also self-replicating.

By contrast with the inner membranes, the outer membrane of mitochondria appears to include components derived from the endoplasmic reticulum.

On the basis of such characteristics it is possible to separate the various types of membranes into three family groups.

(1) Nuclear membrane
Annulate membranes
Rough endoplasmic reticulum
Smooth endoplasmic reticulum
Outer membrane of mitochondria
Microbodies

(2) Golgi apparatus
Primary lysosomes
Heterophagosomes
Secondary lysosomes
Plasma membrane
Multivesicular bodies

(3) Inner mitochondrial membranes

In all probability, however, such groupings are artificial except at a gross level since, as we shall see in the next section, membrane pieces derived from the breakdown of plasma membranes seem to be incorporated subsequently into either Golgi membranes or those of the endoplasmic reticulum even though these two membranes belong to different 'families'.

5.1 Turnover of membranes

The chorus of the old camp-fire song concerned with the well-being of an ageing quadruped, 'The old grey mare, she ain't what she used to be', contained perhaps more of an element of truth from the point of view of cellular and membrane replacement than doubtless the author recognized. Almost all the components of our bodies are constantly being turned over with the result that, apart from the DNA of cell nuclei, parts of the bone structure and perhaps a few lipids in myelin, we are none of us composed of the same molecules as at this time last year. Cells too are replaced in many tissues. Nerves and muscles are long-lived cells but turnover in some other tissues is relatively rapid. Thus human red blood cells survive on average for about 130 days, liver cell life is 160–400 days and that of the epithelial cells forming the epithelial lining of the small intestine is about $1\frac{1}{2}$–4 days. Replacement of dead cells would necessitate a considerable rate of formation of new membranes even if there were no turnover of organelles within the individual cells. In fact the membranes of cells are of course by no means stable; liver mitochondria are destroyed at the rate of about 1 per cell every 15 minutes, half the protein of the endoplasmic reticulum of liver cells (rat) is replaced every 90 hours and the replacement time for lipid is slightly faster still.

The turnover of membranes has been studied by examining the rate of incorporation of radioactive phosphorus, ^{32}P, into the newly formed phospholipids in the cytomembranes of pancreas cells. In short-term experiments most of the incorporated tracer is in the form of phosphatidyl inositol with a substantial, though smaller, amount in phosphatidyl ethanolamine (cf. also Table 6). The dominant phospholipid, phosphatidyl choline, contains little tracer and thus must be turned over more slowly than the other two. This evidence lends support for a theory which suggests that there is a recycling of membrane components within cells and that some parts of old membrane are re-used without the components being synthesized anew. G. E. Palade and his colleagues have shown that α-chymotrypsinogen, the precursor of pancreatic digestive enzyme, is formed in the ribosomes of the rough endoplasmic reticulum and is then transferred first to the cisternae of the rough endoplasmic reticulum and thence, via vesicles budded from the reticulum, to the region of the Golgi. The vesicles coalesce in the Golgi region forming zymogen granules and are then transported to the cell surface. At the surface, the vesicle membrane fuses with the plasma membrane and the enzymic content is released. As a result of these various processes the amount of membranous material in the plasma membrane has been increased by the addition of the vesicle membrane and corresponding decreases in membrane area will have occurred in the Golgi and reticulum. To compensate for this an equivalent amount of membrane, perhaps the added region itself, is therefore degraded into sub-units and removed from the plasma membrane. Since the turnover of lecithin is slow by comparison with that of phosphatidyl inositol it has been suggested by L. E. Hokin that the former may be associated with protein to form relatively stable units in the membrane, whilst the latter has primarily a linking role. In this concept the part played by phosphatidyl inositol would be that of the cement in a membranous wall of lecithin-protein bricks. Once released from the plasma membrane the lecithin-protein units are thought to be transferred through the cytoplasm and reincorporated into Golgi and endoplasmic reticulum membranes to replace the membrane utilized in packaging the secretory enzymes in the first place. The observed synthesis of phosphatidyl inositol which accompanies this process would then represent the recementing of the orginal bricks into newly-formed membrane (Fig. 5–1).

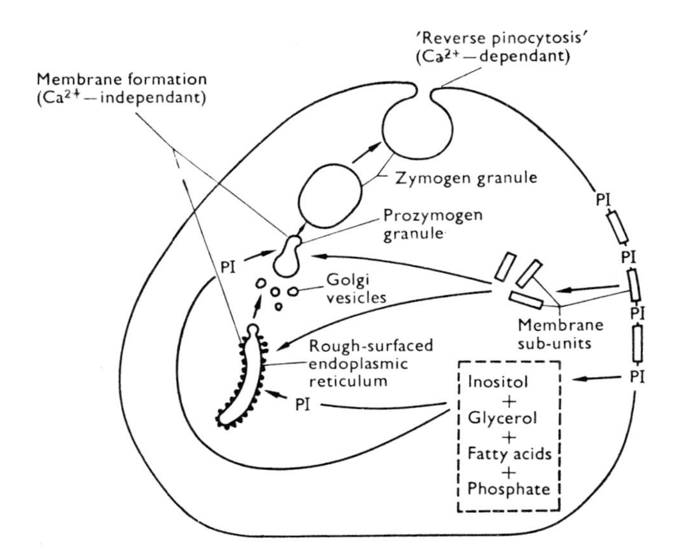

Fig. 5–1 Diagram illustrating the hypothesis of membrane circulation by sub-unit relocation. Membrane added to the plasma membrane by the coalescence of the membranes of zymogen granules (cf. Fig. 4–10) is released by breakdown into sub-units. These are then presumed to be reassembled to form membrane which is used to replace that lost by the rough endoplasmic reticulum and Golgi apparatus in the formation of zymogen granule vesicles. Phosphatidyl inositol (PI) may be the agent responsible for linking sub-units to reform membranes. (From HOKIN, L. E. (1968). *Int. Rev. Cytol.*, 23.)

Certainly, as required by this theory, the fatty acid components of phospholipids can be re-utilized more than once since it has been found that the half time of exchange of the fatty acids of endoplasmic reticulum is longer than that of the polar end-groups of the phospholipids.

The process of turnover in the plasma membrane implied above has been strikingly demonstrated in *Amoeba*. A fluorescent antibody can be prepared which is adsorbed onto the cell membrane and remains readily visible. However, as a result of incorporation of new material and replacement of the original membrane, all the fluorescent material is transferred into the interior of the cell within 24 hours.

Since the amount of a membrane system is determined by the dynamic balance between the rate of formation and rate of breakdown it is possible by adjustments to these rates to alter either the quantity or the properties of a given membrane. Such an effect is seen in cells treated with the hormone cortisone since this steroid tends to cause reduction in the area of the SER, RER and nuclear envelope.

6 Passage of Solutes Across Membranes

One of the most fascinating properties of biological membranes is their ability to restrict the passage of some substances and to permit, or even assist, the movement of others. The lipid layers of cell membranes (p. 6) are a major factor in restricting the permeability and it has been known for many years that substances which are lipid soluble tend to pass cell membranes more readily than those which are water soluble. For example, the lipid-soluble butyric acid penetrates into cells faster than hydrochloric acid of equivalent concentration.

Nevertheless many water-soluble substances do penetrate membranes either by passing along pores through the lipid barrier or by being actively transported by mechanisms forming part of the membrane structure. There are six major ways by which non-lipid soluble materials can cross membranes: (1) diffusion down the concentration gradient; (2) diffusion down an electrical gradient; (3) active transport; (4) exchange diffusion; (5) bulk flow; (6) pinocytosis.

The various ways in which water itself crosses membranes will be considered later.

6.1 Diffusion down concentration gradients

All solutes tend to diffuse down a concentration gradient if not influenced by other factors, and will pass similarly across membranes from a region of high concentration to one of low concentration. Diffusion of oxygen across the epithelial membranes of the lung and across cell membranes is one obvious example of this process.

In biological systems it is sometimes important that the net passage of material across a membrane should be kept high and mechanisms are present to ensure that the diffusing substance is not allowed to accumulate once it has crossed the membrane. A classic example of this can be seen in the passage of glucose from the gut to the blood in insects. J. E. Treherne has shown that glucose passes from the gut lumen to the blood in locusts by diffusion. The concentration gradient is maintained by the conversion of glucose into the disaccharide trehalose by the fat body. Glucose can therefore continue to penetrate into the blood whilst the larger disaccharide cannot escape back from the blood to the gut.

6.2 Diffusion down electrical gradients

Charged materials are influenced by electrical fields and may be caused to diffuse against their concentration gradients when exposed to a sufficiently large gradient of potential. Since potential differences of the order of a few to about 100 mV (millivolts) are found across most epithelial and cellular membranes, this factor is of considerable importance in governing the distribution of charged substances, particularly inorganic ions.

A balance point between the effects of a concentration gradient and a potential gradient on an ion is reached when the following equation is satisfied:

$$zFE = RT \ln (C_1/C_2)$$

where z is the valency of the ion, E is the electrical potential difference across the membrane, F is the faraday (96 500 coulombs/g equivalent), R is the gas content, T is the absolute temperature in °C and C_1 and C_2 are the concentrations of the ion on the two sides of the membrane.*

Considerable differences in the concentrations of certain ions are maintained across membranes by the potential difference. A potential difference of approximately 58 mV with the correct sign will balance a concentration difference of 10 : 1 for a monovalent ion. The chloride concentration of many cells is apparently determined in this way, chloride diffusing down the concentration gradient being balanced by outward diffusion of chloride down the potential gradient across the cell membrane.

6.3 Active transport

It is probably not too wide an assertion to say that the cell membrane of all living cells has the capacity to transport at least one substance by means of active metabolic processes. Active transport in this case is taken to imply that the membrane is capable of transporting the substance in the opposite direction to that in which it would diffuse if influenced only by the electrical and concentration gradients considered above. Work has to be done in transporting materials against the electrochemical gradient and the membrane derives the necessary energy from metabolic processes.

A very considerable number of substances are known to be actively transported by at least some membranes but those which have been most studied are the inorganic ions, amino acids and carbohydrates.

6.3.1 Active transport of sodium

Many of the biologically important inorganic ions such as calcium, magnesium, chloride, potassium, phosphate, etc., are actively trans-

* The letters ln indicate that the value of C_1/C_2 is to be expressed as a natural logarithm. The numerical value of this is 2.303 times the $\log_{10}$ value.

ported by specialized membranes such as those of the gut and excretory organs but only sodium appears to be universally transported by all animal cell membranes. All cells which have been studied have the capacity to extrude sodium, though the ability of cell membranes to do this is poorly developed in the case of those cells such as cat erythrocytes (RBCs) which do not maintain a high internal potassium and low internal sodium concentration. The reason for the universality of the ability of cell membranes to transport sodium is probably associated evolutionarily with the need to regulate cell volume. If a hypothetical artificial cell is surrounded by a membrane which has a finite permeability to inorganic ions and water and is bathed by an isotonic salt solution there will be a tendency for sodium and chloride ions to diffuse in down the concentration gradient. Ionic balance will be reached when:

$$\frac{[Na]_i}{[Na]_o} = \frac{[Cl]_o}{[Cl]_i}$$

where Na_i and Cl_i are the sodium and chloride concentrations inside the cell and Cl_o and Na_o are the outside concentrations. However, if this model cell contains substances, such as protein, which cannot penetrate across the membrane then the osmotic concentration inside the cell at the ionic equilibrium point exceeds that in the external solution. To prevent continued water entry, the membrane must either be sufficiently strong to resist the build up of an internal hydrostatic pressure large enough to prevent further net entry of water or alternatively must have some other means of stopping water coming in. Presumably this same problem was encountered by the earliest forms of life in the sea and it seems likely that the development of the mechanism for the extrusion of sodium forms the basis of the means by which volume can be regulated. According to a current view the way in which many modern cells control their volume by this means is basically as follows.

Sodium diffusing into the cell is actively transported out by the cell membrane. The extrusion of a positively charged ion in this way has to be balanced by the uptake of a similar charge and is in fact achieved by the uptake of potassium ions into the cell. The potassium concentration is maintained at a high level inside the cell and there is thus a tendency for potassium to diffuse out down the concentration gradient. However, since the main anion of the cell is protein which cannot escape across the membrane any loss of potassium creates a separation of charge and hence a potential difference across the membrane. This potential reaches a value such that the electrochemical gradient for potassium across the membrane is zero and there is then no further net loss of potassium (Fig. 6–1). The potential difference across the membrane controls the distribution of chloride ions across the membrane so that:

$$E = \frac{RT}{zF} \ln(\mathrm{Cl_i}/\mathrm{Cl_o})$$

where R is the gas constant, T the absolute temperature, E is the membrane potential and $\mathrm{Cl_i}$ and $\mathrm{Cl_o}$ are internal and external chloride concentrations. As a result of these various effects, all due ultimately to the operation of the sodium extrusion mechanism in the cell membrane, the cell is brought into osmotic balance with its bathing medium.*

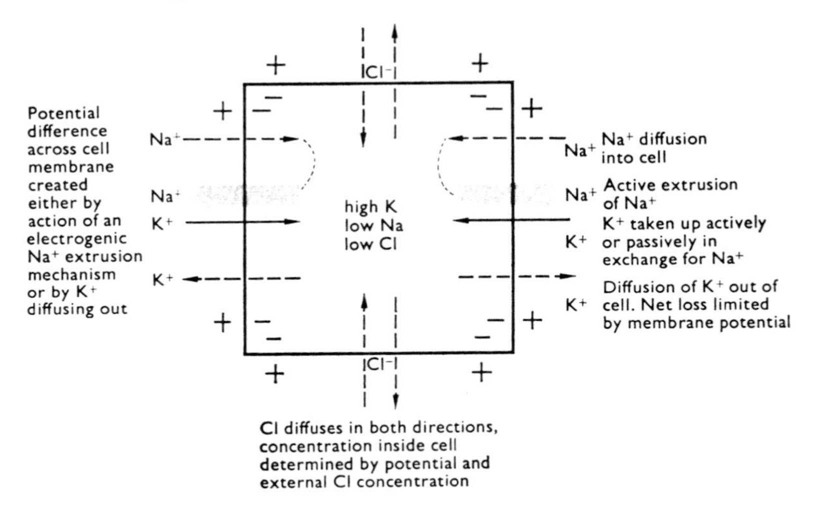

Fig. 6–1 Ion movements across the plasma membrane of cells which are responsible for the production of the membrane potential and regulation of cell water content.

Interference with the operation of the sodium pump results in a disturbance in cellular water content. Thus after cooling to near freezing red blood cells increase their water content and sodium content and there is a loss of potassium.

Rewarming the cells to blood temperature in the presence of metabolite, glucose, results in the extrusion of the excess sodium and recovery of the lost potassium. Much of the excess water is also removed.

The precise method by which sodium is actively transported by cell membranes is still far from clear. Certain features of the process have been established, however: (1) the mechanism is energy consuming and uses adenosine triphosphate (ATP) as its energy source, (2) there is a direct relationship between the energy utilization and the number of sodium

* This explanation of cellular osmotic balance is of course oversimplified. Much controversy remains in this field and for an alternative view see ROBINSON, J. R. (1965). Water regulation in mammalian cells. *Symp. Soc. exp. Biol.*, **19**, 237.

ions transported, (3) an ATP hydrolysing enzyme specifically activated by sodium and potassium plays a fundamental role in the process, (4) the rate of transport of sodium is decreased if the potassium concentration of the medium is lowered or if the ratio of potassium to sodium inside the cell is increased.

Adenosine triphosphate appears to be the prime, or perhaps even the sole, source of energy for sodium transport and even closely related compounds with high energy phosphate bonds cannot be utilized. For each ATP molecule split 3 sodium ions are transported across the membrane of red blood cells, a value which agrees well with the value of 2–3 ions per phosphate bond obtained for sodium transport by the frog skin. Work on the frog skin and other tissues indicates that the relationship between sodium transport and energy utilization is constant irrespective of the electrochemical gradient against which the sodium is being moved. This is contrary to the result which we would expect from thermodynamic considerations since the minimum work to be done should vary with the gradient according to the equation:

$$W = \boldsymbol{RT} \ln([\mathrm{Na_i}/\mathrm{Na_o}]) + z\boldsymbol{FE}$$

where W is the work and other symbols are as previously defined. As there is no such relationship to the gradient it must be presumed that the efficiency of the process varies with the electrochemical gradient, being most efficient when there is a large gradient.

The enzyme Na–K–Mg activated ATPase has been isolated from the cell membranes of red blood cells and squid nerve axons both of which are sites where sodium transport occurs. Certain races of sheep which have very little of the enzyme present in the red blood cell membrane have blood cells which are atypical in having a high internal sodium concentration. By contrast other races of sheep which have blood cells with a low sodium content have more of the enzyme in the red cell membrane. Similarly the high sodium blood cells of cats have less enzyme in their membranes than the low sodium blood cells of man. The most logical explanation of these differences is that the enzyme is engaged in the provision of energy for sodium extrusion from cells.

The transport of ions across an epithelial membrane such as a gill, gut wall or excretory tubule wall involves a number of features not present in transport across a single cell membrane. Of major importance is the fact that the ions have to be moved into the epithelial cell across one cell membrane and out across the other (Fig. 6–2). The inner and outer cell membranes must therefore differ in their properties. Another difference is that an epithelium must transport sodium by a means which is not dependent on an exchange of sodium for potassium as in the case of the cell membrane. Finally, epithelia at the body surface must have the capability of transporting chloride actively as well as sodium. Since

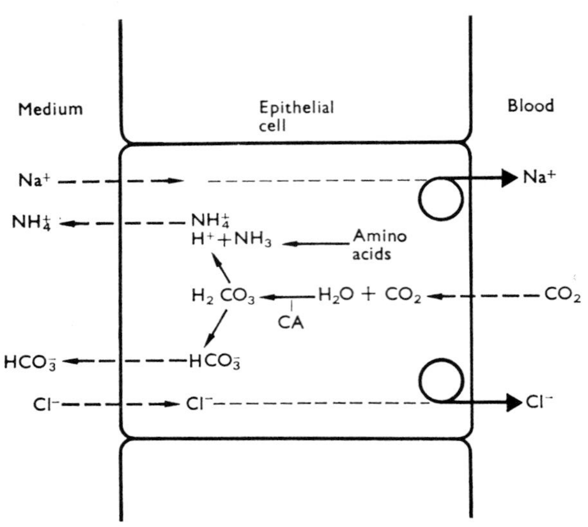

Fig. 6–2 Processes involved in uptake of NaCl by the gills of freshwater fish. Sodium uptake occurs in exchange for NH_4^+ ions derived in part from the deamination of amino acids. Chloride is exchanged for HCO_3^- derived from the ionization of H_2CO_3. Na and Cl are actively transported across the inner boundary of the cell into the blood. (Simplified from MAETZ, J. and ROMEU, GARCIA F. (1964). *J. Gen. Physiol.*, **47**.)

animals tend to lose sodium and chloride at different rates the mechanisms for sodium and chloride uptake must be independent of one another. Independent transport of sodium and chloride at the body surface has been demonstrated on fish, frog skin and crustacean gills. Crayfish, for example, can take up chloride from NaCl, KCl, $CaCl_2$ and NH_4Cl and sodium from NaCl, $NaNO_3$, Na_2SO_4 and $NaHCO_3$. Except in the case of sodium chloride the other ion is not taken up. Now, as in the case of a single cell, uptake of an ion by the surface epithelium can only occur if an ion of opposite charge is also taken up or if an ion of the same charge is lost from the body. J. Shaw has shown that, in the crayfish, sodium uptake is largely accounted for by exchange with ammonium and hydrogen ions whilst chloride is exchanged for bicarbonate ions.

A similar exchange of sodium for ammonium and chloride for bicarbonate occurs in the goldfish. The enzyme carbonic anhydrase (C.A.), which catalyses the reaction is clearly of considerable importance in the

$$H_2O + CO_2 \xrightleftharpoons{\text{C.A.}} H_2CO_3 \rightleftharpoons H^+ + HCO_3^-$$

production of the ammonium and bicarbonate ions involved in the exchange because when it is inhibited by the injection of acetazolamide both sodium and chloride transport are diminished. The hypothetical role of carbonic anhydrase is indicated in Fig. 6–2. Carbon dioxide diffuses into the epithelial cells from the blood and is converted to H_2CO_3 by the enzyme. On ionization this produces the bicarbonate necessary for chloride exchange and hydrogen ions. The hydrogen combines with ammonia produced by the breakdown of amino acids to give ammonium ions for exchange with sodium. It seems likely, however, that a primary Na^+/H^+ exchange can occur when ammonia is not available. Sodium and chloride ions, once inside the cell, are moved across the inner cell membranes by active transport processes.

6.3.2 Factors affecting the rate of sodium transport

Various factors influence the rate at which sodium is transported by epithelia. These are of two types, intrinsic, i.e. factors arising within the body, and extrinsic where the cause is external. The principal intrinsic factor is control by hormones. Variations in blood sodium concentration or blood volume are followed by changes in the amount of hormone circulating in the blood. This in turn determines the activity of epithelia at the body surface such as gills and renal tubules. In vertebrates the hormone which has the greatest effect on sodium transport is the steroid, aldosterone.

Aldosterone seems to act by stimulating the production of new protein by transporting cells. The nature of the protein formed is unclear but it seems reasonable to suppose that it is one of the components of the sodium transporting mechanism in the membrane. In other words an animal's response to ion imbalance is to vary the number of transporting sites. Whether this occurs during routine replacement of membrane (p. 45) or by specific incorporation in existing membrane is not known. No comparable hormone has yet been isolated from invertebrates.

By contrast, extrinsic factors which alter the rate of sodium transport appear to do so by affecting the rate of transport by the existing sites. The most important extrinsic factors are temperature, sodium concentration, pH and ammonium concentration.

Sodium transport involves chemical reactions and, like any other chemical reaction, is affected by the temperature. Only mammals and birds are unaffected by this because they maintain a constant body temperature. In all 'cold-blooded' forms (poikilotherms) the rate of transport is increased by a rise in ambient temperature within the physiological range of the animal.

The concentration of sodium in the medium affects the rate of transport only at low levels. For instance if the concentration in the medium is lowered there is no change in the rate of uptake of sodium by a crayfish until the sodium level has fallen to about 1 mEq./l. (23 mg/l.). At concentrations below this the maximum rate of uptake of which the animal is capable is related to the concentration of the medium, declining more rapidly the greater the dilution. The shape of the curve (Fig. 6–3)

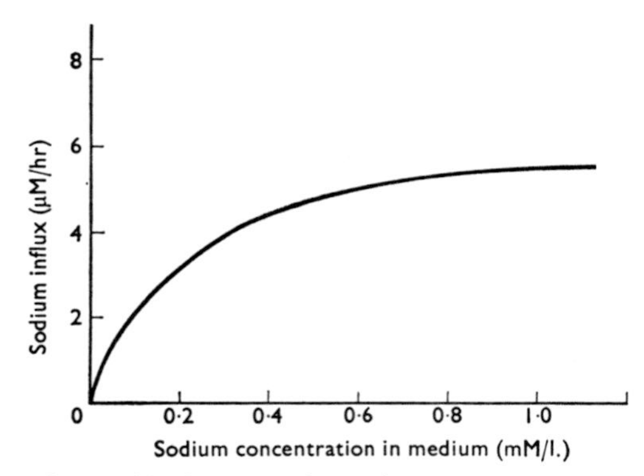

Fig. 6–3 The relationship between the sodium concentration of the medium and sodium influx into crayfish. (Simplified from SHAW, J. (1959). *J. exp. Biol.*, **36**.)

which shows the relationship between sodium uptake and concentration of sodium in the medium is a hyperbola. Such a shape would be expected if a carrier mechanism ferried sodium across one of the epithelial membranes. It is assumed that at low concentrations of sodium in the medium this carrier is not fully saturated.

The concentration required to produce half the maximal rate of sodium transport (K_m) varies widely in different animals. Some, such as the shore crab *Carcinus*, need a fairly high concentration of sodium in their medium to saturate their transport sites. Species living in more dilute media have transport mechanisms which are capable of their maximum rate of transport at much lower concentration (Table 7).

Table 7 The concentration of sodium in the medium required to give half the maximal rate of sodium transport in crustacea from different habitats.

Species	*Habitat*	K_m (mE/Na)
Marinogammarus marinus	Marine	6–10
Gammarus duebeni	Brackish water	1.5–2.0
Gammarus pulex	Fresh water	0.1–0.15
Sphaeroma serratum	Marine	15
Sphaeroma rugicauda	Brackish water	2.3
Mesidotea entomon	Brackish water (Baltic)	12
Mesidotea entomon	Fresh water (Vättern)	1.23

It is clear from this table that, during the course of their evolution from marine ancestors, freshwater forms must have evolved modifications to the transporting system which make the mechanism able to work effectively at low concentrations of sodium.

Increasing either the hydrogen ion concentration (lowering the pH) or the ammonium concentration in the medium tends to decrease the active uptake of sodium. Probably this is due to competitive interference with the sodium-ammonium exchange process.

6.3.3 *Transport of organic substances*

Numerous organic materials including such substances as glucose and other monosaccharides, amino acids, pyrimidines, bile salt, phenol red, penicillin, etc., can all be transferred by at least some membranes against an apparent concentration gradient. Many of the transfer mechanisms have certain features in common implying that the principle of the processes may be similar. Most striking is the fact that the presence of sodium is essential to the movement of glucose and other monosaccharides, amino acids, pyrimidines and possibly the bile salts. In order for transport to occur, sodium must be present at a higher concentration on the side of the cell membrane from which the organic material is to be moved than on the other side. A high concentration of potassium, lithium or ammonium ions in the medium outside the cell interferes with transfer of organic material into the cell.

The interpretation placed on these and other findings is that organic materials cross cell membranes bound to a carrier molecule which also binds sodium. The system was originally presumed to operate as shown in Fig. 6–4. In this hypothesis, the carrier substance, probably a protein, has two active sites which are temporarily exposed at the outer surface of the cell membrane. A sodium ion becomes associated with one of the sites and this results in the carrier assuming a configuration suitable for the adsorption of an organic molecule, say glucose, at the other site. Random (thermal) movement now transfers the carrier plus the associated glucose and sodium to the inner face of the cell membrane. The sodium concentration inside the cell is low and the potassium concentration is high so that the sodium ion attached to the carrier tends to exchange for a potassium ion. This exchange alters the configuration of the carrier again so that the glucose is released inside the cell. The carrier is then carried back by thermal movement to its starting point and the cycle is repeated. The overall effect is to transfer glucose and sodium into the cell and potassium out.

In the light of recent findings that interchange of protein molecules from one surface of a membrane to the other is a rare event but that some proteins are so located as to traverse the membrane, the details of Crane's hypothesis require revision. The principal of his suggestion remains valid however provided that the inorganic ions produce configurational changes in a membrane-traversing protein such that forced passage of glucose across the membrane is facilitated.

The operation of such a system will clearly be interfered with if there is a high concentration of potassium outside and a high concentration of sodium inside the cell. Indeed when these circumstances are artificially

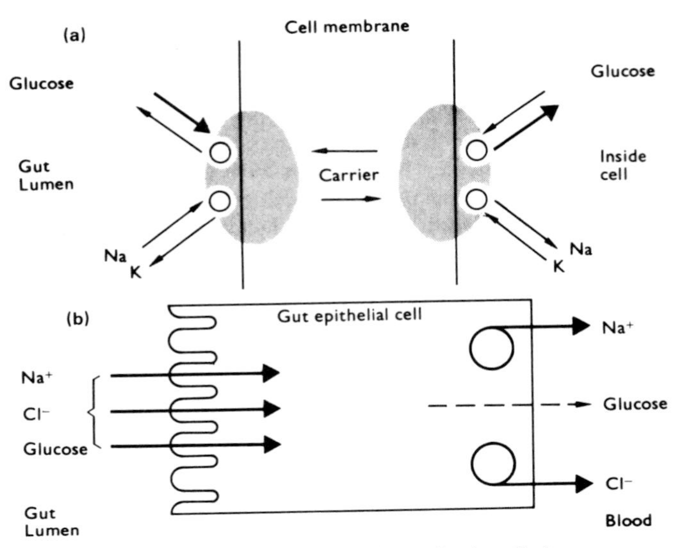

Fig. 6–4 Diagram to illustrate Crane's hypothesis of the manner of the relationship between ion movements and the translocation of glucose (or amino acids) across the gut epithelium. (**a**) Adsorption of sodium onto a carrier molecule alters its spatial configuration so that a glucose can be absorbed. After passage of the carrier across the membrane Na may be lost in exchange for K and the glucose is then liberated inside the cell. (**b**) Glucose inside the cell diffuses down the concentration gradient into the blood. (Modified from R. K. CRANE (1965). *Fedn. Proc. Fedn. Am. Socs exp. Biol.*, 24.)

produced in gut cells, the direction of glucose movement is reversed and it can be transferred into the gut against a concentration gradient. In a living gut cell it is likely that glucose does not accumulate but will rather diffuse out of the cell into the blood across the opposite membrane to that by which it entered. Appropriately it appears that little of the sugar transported is available to the metabolic processes of the transporting cell.

Different carrier molecules are probably involved in transferring other organic materials though the underlying process may be similar to that outlined for glucose. It appears likely that three or four different carriers are responsible for transferring amino acids since amino acids can be separated into groups according to whether or not they compete with each other for transfer. Members of a group compete with other members of the same group but not with members of other groups.

6.4 Exchange diffusion

Exchange diffusion is an obscure process by which an ion on one side of a membrane is exchanged for an ion of the same chemical type on the other side. The fact that this exchange occurs at all therefore can only be recognized if the transfer of radioactive ions across a membrane is studied. The membrane cannot distinguish between radioactive and non-radioactive ions and so both types are transferred with equal facility.

The higher the concentration of the appropriate ion in the medium the greater is the chance that an exchange will be effected. Ion exchange diffusion can only occur at a fast rate therefore when there is a concentrated medium on both sides of the membrane. Thus transfer of an euryhaline fish from saline to fresh water is generally accompanied by a simultaneous decline in the efflux across the gill of both Cl^- and Na^+. Similarly transfer of *Artemia salina* from sea-water to a non-ionic medium isotonic with sea-water results in a drastic decline in Na^+ efflux (Fig. 6–5). Two types of mechanism may be involved in such instantaneous change in ion flux: (1) alteration of the electrical potential difference across the surface epithelium; (2) involvement of a carrier system in the exchange

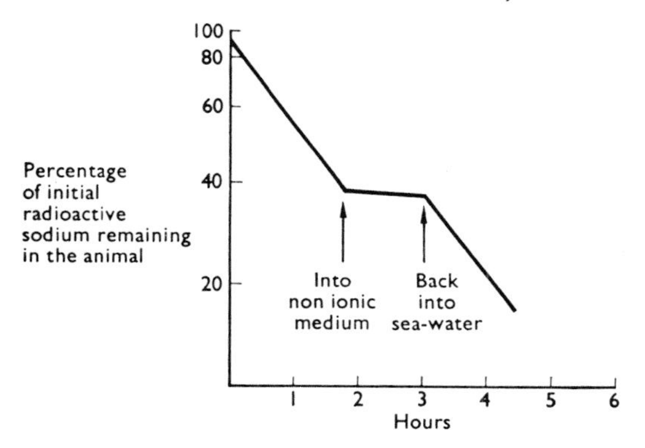

Fig. 6–5 The loss of labelled sodium from *Artemia* to sea-water and to a non-ionic medium isotonic with sea-water. Note that the loss rate declines in the non-ionic medium, implying that exchange diffusion contributes to the loss to sea-water. (Simplified from CROGHAN, P. C. (1958). *J. exp. Biol.*, 35.)

process. In sea-water the inside of a fish gill is positive to the outside but on transfer to fresh water the direction of the electrical gradient may change so that the outside becomes positive. As a consequence it would be expected that following dilution of the medium the tendency for sodium ions to move outward down the electrical gradient would diminish. Such an explanation could not, however, explain the contemporaneous decline in Cl^- efflux since, for this ion, the gradient should increase on dilution of the medium. It seems necessary therefore to assume that at least in the case of chloride a major part of the exchange flux involves a carrier mechanism. This view receives added support from the fact that much of the passive efflux of chloride to sea-water, as well as the actively extruded component, can be blocked by SCN^-.

Since exchange diffusion involves a 1 : 1 exchange of an ion across the membrane there is in theory no performance of work and no energy is required to effect the exchange. It is doubtful if exchange diffusion is of any physiological significance to animals though it is a nuisance to experimenters attempting to evaluate the active movement of ions.

6.5 Bulk flow

If a membrane is permeable to substances other than water they may traverse the barrier in water moving under the influence of hydrostatic or osmotic forces. Perhaps the best known examples of such bulk flow is the formation of primary urine in the vertebrate kidney by hydrostatic pressure filtration.

The hydrostatic pressure of blood in the glomerulus of the kidney exceeds the colloid osmotic pressure of the plasma. An ultrafiltrate of blood is therefore forced across the permeable wall of Bowman's capsule into the lumen of the excretory tubule.

6.6 Pinocytosis

This is the name given to a process by which cells engulf a portion of their medium and subsequently absorb the enclosed material. Different forms of pinocytosis are found in amoebae and the cells of higher organisms.

It is obvious that simple engulfing of the medium cannot *by itself* be of value in *selectively* transporting material from the medium into a cell because all that is achieved is a transfer of medium, including everything it contains, into the cytoplasm. However, pinocytosis is rather more complex than a simple engulfing of fluid. If the protein which stimulates the onset of pinocytosis is labelled with dye or radioactive material it is found that the label is adsorbed onto the surface of the cell (or amoeba)

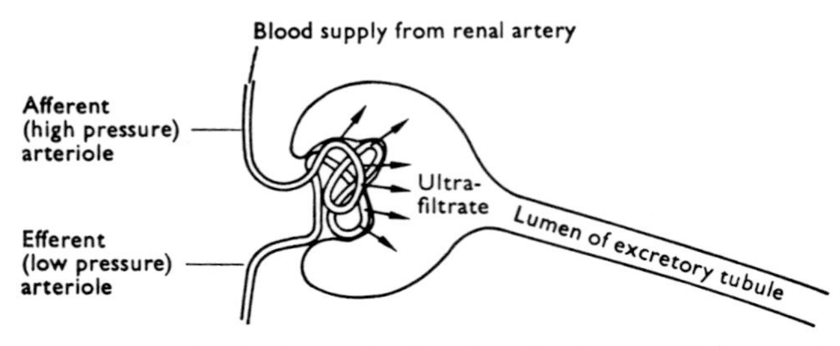

Fig. 6–6 Bulk flow by hydrostatic filtration in the vertebrate nephron. An ultrafiltrate of blood passes from the glomerulus into the lumen of the nephron.

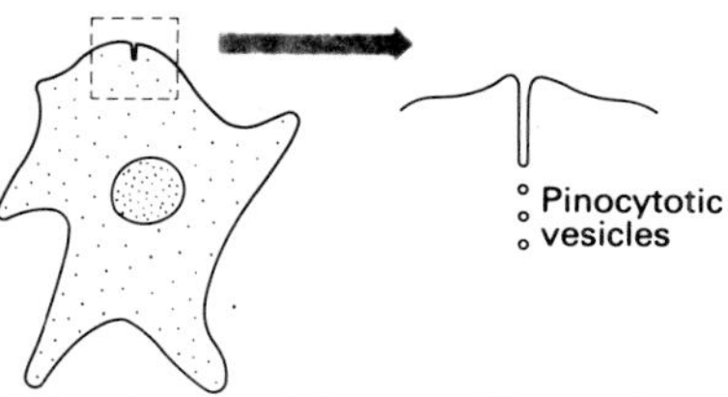

Fig. 6–7 Pinocytosis in *Amoeba.* Diagram illustrating the formation of an intucking of a portion of the surface with pinocytotic vesicles being budded off at the bottom of the canal so formed.

and that a higher concentration of label subsequently appears in the vesicles than is present in the bulk of the medium. This implies that the formation of the pinocytotic vesicles is a means of taking into the cytoplasm portions of the cell membrane which have selectively adsorbed material from the medium.

It is not yet clear if the part of the surface which is involved in adsorption and formation of pinocytotic channels is specialized but it may be significant that after about half an hour in a protein solution pinocytosis by amoebae ceases and cannot begin again for several hours. This could be interpreted as indicating that a specialized part of the membrane has been used up and has to be replaced before adsorption can continue.

Another difference between the process in amoebae and tissue cells is that it cannot readily be induced in the latter by the presence of protein. Possibly special inducers are required by tissue cells. For a discussion of the fate of pinocytotic vesicles within cells see p. 35.

An analogous process to pinocytosis is sometimes used to remove substances from cells. The contractile vacuoles of amoebae are formed by the coalescing of preformed small vacuoles with the membrane of the main vacuole. The membrane of the main vacuole ultimately fuses with the cell membrane and the vacuole ruptures liberating its contents. In this example there is no loss of membrane. By contrast, secretory cells sometimes extrude vacuoles as a means of liberating their secretion without loss of other cellular components (cf. p. 36). Such a mechanism is found in pancreatic acinar cells. A comparable production of vesicles occurs in the excretory tubules of many animals. The crayfish provides an example of cells which show both pinocytosis and vesicle formation. Pinocytosis occurs at the border of the excretory organ cells next to the blood and the excretory organ cells also form and release vesicles at their border with the lumen of the excretory tubule. These cells therefore provide a possible means of removing from the blood any substance which cannot readily enter the excretory tubule by the usual filtration process. The vesicles released into the tubule lumen contain proteolytic enzymes which may render some of the contents suitable for reabsorption by the bladder wall.

7 The Movement of Water Across Membranes

7.1 Diffusion and osmosis

Even when the concentration of water on the two sides of a permeable or semi-permeable membrane is the same water exchanges rapidly in both directions by diffusion. This process can readily be observed if some tritiated (radioactive) water is added to the solution bathing one surface of the membrane. The labelled water exchanges across the membrane so that ultimately the ratio of tritiated water to ordinary water is the same on both sides of the membrane. When no other forces are exerted the total amount of water on each side of the membrane remains constant since the diffusion between two solutions of the same water concentration is equal and opposite.

If the solutions separated by the membrane differ in concentration initially then diffusion of water in the two directions will be unequal and there will be a net transfer of water across the membrane so that the volume of the more dilute solution will be decreased and that of the more concentrated increased. This process is termed *osmosis*. In an artificial system net transfer of water will cease, when the hydrostatic pressure increase, resulting from the gain in volume of one solution at the expense of the other, balances the difference in osmotic pressure of the two solutions. Theoretically, the osmotic pressure of a one *molal* solution of any non-electrolyte will be balanced by an hydrostatic pressure difference of 22.4 atmospheres at 0°C (approximately 230 m of water). At higher temperatures a larger hydrostatic pressure is necessary to achieve balance since osmotic pressure increases with the absolute temperature according to Charles' law.

$$P_{t_1} = P_{t_2}\left(1 + \tfrac{1}{273}(t_1 - t_2)\right)$$

where P_{t_1} and P_{t_2} are the osmotic pressures at the temperatures t_1°C and t_2°C respectively.

Theoretically, solutions of non-electrolytes having equal *molalities* have the same osmotic pressure at any given temperature. In practice, however, there are small differences and in calculations of osmotic pressure from concentration a correction factor, the *osmotic coefficient*, must be included.

It must be stressed that it is solutions with equal *molalities*, not those of equal *molarities*, whose osmotic pressures tend toward equality. *Molal* and *Molar* solutions differ, of course, in the fact that in the case of a 1 molal solution the molecular weight of a substance in grammes is dissolved in 1 000 g of water whereas for a 1 molar solution the molecular weight in grammes is dissolved in water and the solution is then made up to 1 000 ml. with water.

7.2 Osmosis in biological systems

Cellular membranes have a finite permeability so water tends to be distributed between the major fluid compartments of the bodies of animals according to the solute content of the compartments. For example, consider four conditions affecting the distribution of water between the cellular and extracellular compartment in a mammal. These are: (1) an excess of water present in the body, (2) a water deficit, (3) excess salt in the body, (4) a salt deficit.

(1) Excess water is distributed between extra- and intracellular compartments so that their relative concentrations and volumes remain unchanged.

(2) Similarly, when there is a water deficit there is no change in the relative volumes of cells and extracellular fluid as water is lost proportionately by both.

(3) Sodium chloride in the body is largely extracellular and so when an excess of salt is present this tends to increase the solute content of the extracellular space without grossly increasing that of the intracellular space. Consequently water is withdrawn from the cells by osmosis, thus increasing the extracellular volume at the expense of the intracellular volume.

(4) Conversely, when there is a salt deficit there is a tendency for water to move from the extracellular space to the cells.

In normal circumstances it is the function of the regulatory systems of the body to ensure that net movements of water between compartments do not occur but in certain situations specific use is made of osmosis across either cellular or epithelial membranes to achieve specialized end-results. Perhaps the best known example of this process is in the production of urine more concentrated than the blood by the kidneys of birds and mammals. As another example of the utilization of osmosis we may consider the manner in which the cuttlefish, *Sepia*, balances osmotic forces against hydrostatic forces in order to regulate its buoyancy.

The familiar cuttlebone of *Sepia* is composed of a number of narrow chambers, about 100 in the adult, separated from each other and each bounded on three sides by calcified chitin (Fig. 7–1). The lower end of the chambers is closed by an epithelial membrane. The chambers are filled partly by gas which has diffused in and partly by fluid. The buoyancy of the cuttlebone is such that it just balances the tendency of the other tissues of the body to sink in sea-water and its presence thus gives the animal as a

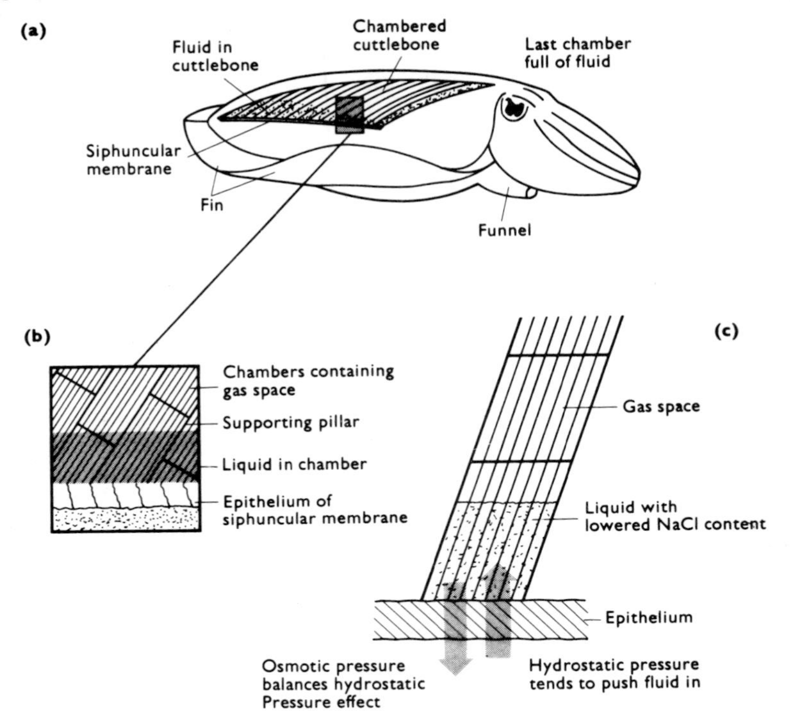

Fig. 7–1 Diagrams illustrating the role of the siphuncular membrane in the buoyancy control of the cuttlefish, *Sepia*. (**a**) Cuttlefish showing the position of the chambered cuttlebone. (**b**) Enlargement of part of the cuttlebone illustrating the chamber sub-structure and position of the siphuncular membrane. (c) Siphuncular membrane controls the fluid volume of the chambers by regulating the NaCl content of the fluid. (Adapted from DENTON, E. J. and GILPIN BROWN, J. B. (1961). *J. mar. Biol. Assoc. U.K.*, **41**.)

whole neutral buoyancy. The animal does not therefore have to expend so much energy in swimming as would be the case if its density were greater than that of sea-water. Hydrostatic pressure increases with depth so when a cuttlefish dives there is a tendency for the increased pressure to force more water into the cuttlebone at the expense of the gas space. This would result in the animal becoming more dense and having to swim harder if it did nothing about it. E. J. Denton has shown, however, that it maintains a constant buoyancy by actively transporting NaCl from the cuttlebone fluid to the blood as it dives deeper. This dilutes the cuttlebone fluid so producing an osmotic gradient across the epithelium. The rate of transport of NaCl is adjusted so that the tendency of the hydrostatic pressure to push water into the cuttlebone is balanced by the opposite tendency for water to move out down the osmotic gradient.

Certain species of toad, including *Bufo cognatus*, which live in environments which are liable to periodic dry spells, have capacious bladders. These bladders can store a volume of urine equivalent to 30% of the total fluid volume of the body. When the animals are exposed to water loss by dehydration anti-diuretic hormone (ADH) is released into their blood stream and renders the bladder wall permeable to water. As the body fluids are more concentrated than the stored urine, water is extracted by osmosis from the bladder thus replacing water lost from the blood by evaporation.

7·3 Isotonic water movement

So far we have considered only systems where there is movement of water under the influence of an osmotic gradient. There are, however, many places within the body where there is a fluid movement across epithelial membranes between solutions of similar or equal osmotic concentrations. Examples include such processes as the secretion and reabsorption of the digestive juices, the secretion of cerebro-spinal fluid by the choroid plexus, the secretion of bile by the liver, the concentration of bile by the gall bladder and the reabsorption of primary urine from the proximal tubule in the kidney. Investigation of these processes indicates that fluid transfer will continue even against a limited osmotic gradient so it is clear that classical osmosis cannot be involved. Theoretically a number of other mechanisms might be responsible for net fluid movement. These include: (1) active transport of water, (2) pinocytosis, (3) pressure filtration, (4) electro-osmosis, (5) co-diffusion, (6) double membrane effect, (7) local osmosis. The assumed role of each of these processes is outlined below.

(1) Active transport of water By definition the active transport of water involves a mechanism whereby water molecules are actively transported across a membrane independently of the active movement of solute molecules. There is no positive evidence that this ever occurs in vertebrates.

(2) Pinocytosis (p. 60) Transport of fluid across a cellular epithelium by this means would involve the engulfing of a quantity of medium in the form of a vesicle, absorption of the fluid and a subsequent secretion of the absorbed fluid across the cell membrane on the opposite side from that on which pinocytosis occurred.

(3) Pressure filtration Under the influence of a hydrostatic pressure gradient fluid will traverse a partially permeable membrane separating two solutions of equal osmotic concentration.

(4) Electro-osmosis It is assumed that pores through a membrane are lined with fixed compounds with free electrical charges. The pore channels will therefore contain mobile ions of opposite charge to balance the fixed charges. When a potential difference is applied across the membrane the mobile ions will move towards the appropriate electrode and in doing so will sweep with them some of the fluid in the pores. There is thus a transport of fluid across the membrane.

(5) Co-diffusion The theory of co-diffusion again involves a system of pores with the water being swept through by the movement of solute particles but in this case the driving force is assumed to be the active transport of the solute by a metabolic process rather than an electrical effect.

(6) Double membrane effect This is a variation on the local osmosis theory presuming that fluid transport is across two membranes of differing permeabilities which lie in series. It is postulated that transport of solute across the first membrane creates a local osmotic gradient so that fluid also traverses the membrane. This fluid movement creates a hydrostatic pressure increase between the membranes which then causes bulk flow of fluid through the more permeable second membrane.

(7) Local osmosis In local osmosis it is assumed that active transport of solute creates an osmotic gradient in the immediate vicinity of the membrane by decreasing the concentration on one face of the membrane and increasing it on the other. Fluid will move across the membrane down the local osmotic gradient. Deep intuckings of the plasma membrane may assist the maintenance of locally raised osmotic concentrations.

To postulate mechanisms by which water might be transported is one thing, to establish the nature of the means actually utilized by animal systems is quite another matter. Attention is currently centred, however, on model systems which take into account not only the kinetics of transfer processes across epithelia but also the detailed morphology of the transporting cells. Epithelial cells which have the capacity to transport ions or water are generally characterized by complex infolding of the lateral and apical plasma membrane (Fig. 2–4). Histochemical staining demonstrates that Na–K–ATPase is associated with the intucked regions as well as with the more superficial membranes. The presence of the ATPase implies that sodium can be transported out of the cell and concentrated in the intercellular channels created by the infoldings. Given that membranes have a finite permeability, water will tend to follow the ions down the concentration gradient. If one end of the intercellular space between neighbouring cells is sealed by a tight junction then ultimately there will be a flow of ions plus water at the other end (Fig. 7–2). In theory, variations in the relative rates of water and ion movement could produce isosmotic or hypertonic fluid transport and forward or backward flow in simple systems (Fig. 7–3) and hypotonic transport in more complex systems. (Fig. 7–4.)

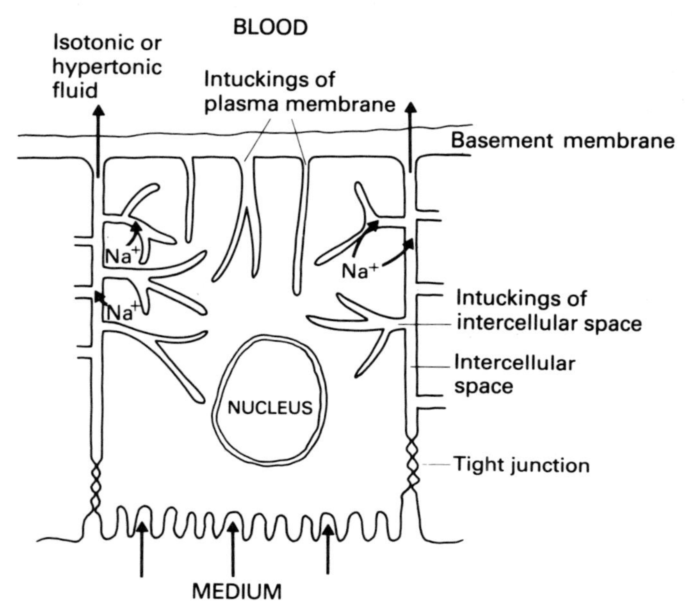

Fig. 7–2 Stylized diagram of epithelial cell responsible for isotonic fluid transfer from media to blood. Intucking of the plasma membrane from both the intercellular channels and the basal surface are shown. Arrows indicate the direction of sodium and water movement. Na–K–ATPase lines the plasma membrane intuckings.

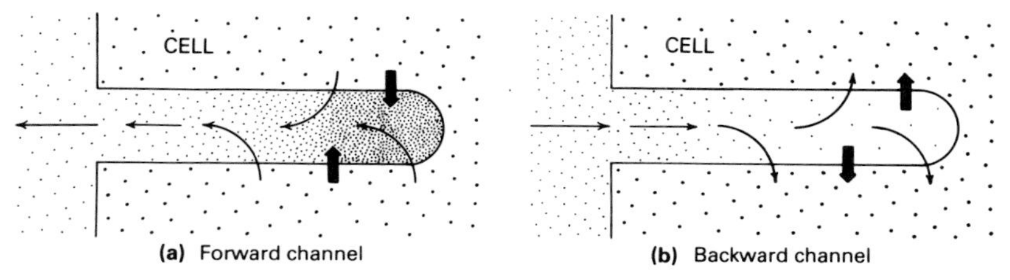

Fig. 7–3 Standing gradient models. (**a**) Forward channel. Ions pumped into channel (broad arrows) create osmotic gradients drawing water from cell (thin arrows). Fluid flows out of open end of channel. (**b**) Backward channel. Ions pumped from channel cause water uptake into cell. Fluid flows into open end of channel. (From OSCHMAN, J. L., WALL, B. J., and GUPTA, B. L. (1974). *Symp. Soc. exp. Biol.*, **28**.)

The rectal pad cells in insects seems to represent an example of this last mentioned process; an ingenious morphological arrangement permitting transfer of a hypotonic solution from the rectal lumen to the blood even in the absence of transportable ions in the rectal fluid. By the use of ultra-microsampling techniques B. J. Wall has been able to extract fluid samples from the intercellular sinuses of rectal pad cells and has shown that the osmotic concentration of this fluid exceeds that of the rectal lumen contents. On the basis of these measurements she and her

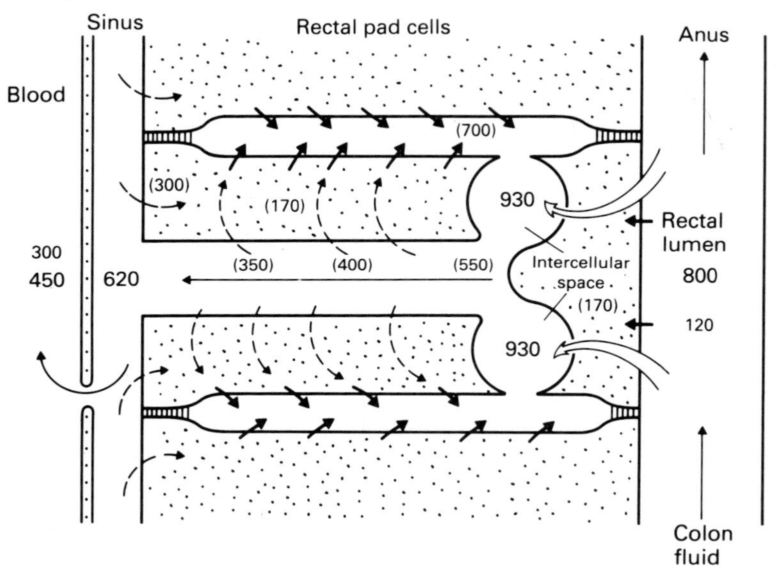

Fig. 7–4 Diagrammatic arrangement of the structure of cockroach rectum to explain the solute-linked absorption of water from the rectal lumen by recycling of solute (Na^+). Bold numbers indicate the average values of measured osmotic concentrations in various compartments. It is postulated that Na^+ and some other organic solutes are pumped into narrow intercellular channels (bold, slanting arrows) to create a very high osmotic concentration in intercellular spaces. Water is then passively drawn (hollow arrows) from the lumen. The fluid moves into the intercellular sinus and ultimately emerges through valves in the muscular sheath (thin arrows). During the latter part, solutes are reabsorbed into the cells (dashed arrows) to refeed the solute pumps. The merging fluid is thus nearly isosmotic with the blood but grossly hypo-osmotic to the rectal lumen. Small numbers in the rectal lumen and in blood are the measured concentrations of Na (mmol/l). Small numbers in parentheses in other compartments are speculated values for Na concentrations. Normally solutes are also absorbed from the rectal lumen, but with solute recycling within the tissue, availability of permeable solutes in the rectal lumen will not be essential for the osmotic absorption of water. (From GUPTA, B. L. (1976). In: *Perspectives in Experimental Biology*, ed. P. S. Davies, Pergamon Press, Oxford and New York.)

colleagues have proposed that hypotonic water transfer is achieved by a system comparable to that illustrated in Fig. 7–4. It is postulated that inorganic ions, supplemented perhaps by organic solutes, are transported from the cell into the intercellular spaces thereby raising the osmotic concentration to a value (930 m. osmols) in excess of the rectal fluid concentration (800 m. osmols). Water is assumed to pass from the rectal lumen into the sinuses and the increased fluid bulk in the latter creates a flow down the intercellular canal towards the blood. During this passage some of the ions and organic solutes are thought to be recovered by active transport processes and subsequently recycled and the remaining fluid, now hypotonic to rectal contents, passes into the haemolymph. Whilst certain components of this model still require to be formally established by experiment, the whole offers an attractive explanation of the means by which some insects can extract the major part of the fluid present in the rectal contents.

Osmoregulation in Hydra. It has been suggested by P. T. Marshall that the regulation of water content of the cells of *Hydra* may also be based upon an ion transport system (Fig. 7–5). The cells of this animal have an internal concentration some 20 m. osmol. in excess of that of the medium and so they tend to take up water by osmosis. Marshall's hypothesis is that inorganic ions are actively transported by the ectoderm cells into the mesogloea and that the endoderm cells then transfer the ions into the enteron until the concentration of the latter just exceeds the cell osmotic

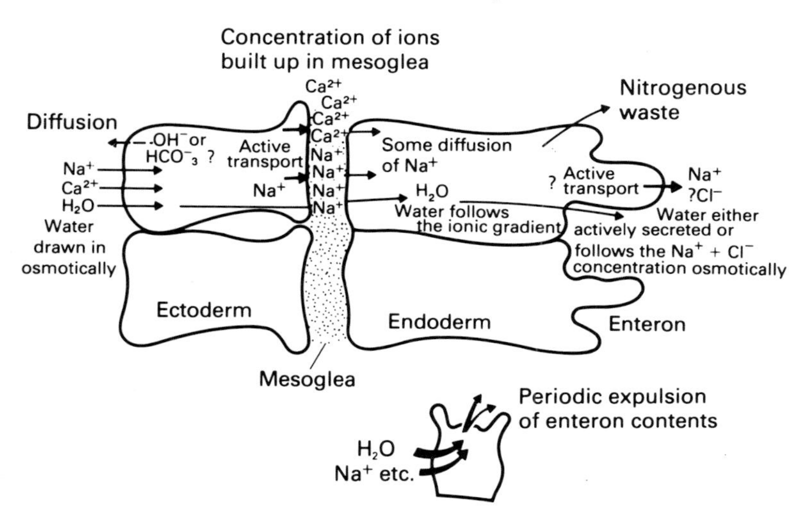

Fig. 7–5 Diagram illustrating the hypothesis that active transport of inorganic ions across the body wall of *Hydra* is the underlying mechanism of volume regulation of the cells. (From MARSHALL, P. T. (1969). *School Science Rev.*, **51**.)

pressure. This is assumed to draw excess water from the cells by osmosis and periodic release of fluid from the enteron disposes of this excess. Evidence in support of the theory is that (1) the animal is known to take up radioactive sodium from its medium, (2) the measured osmotic pressure of the enteron fluid is about the same as that of the cells, (3) fluid is normally released from the enteron about once an hour and (4) ligatured animals swell below the tie due to the accumulation of fluid.

8 Specialization of the Plasma Membrane: Nerves

8.1 Gross structure

Nerve cells consist of a *cell body* containing the nucleus and a long process termed the *axon*, which can be up to several feet in length in large animals. Shorter, branching processes, *dendrites*, extend from the cell body (Fig. 8–1).

In vertebrates most of the axons with diameters exceeding 1 μm (except

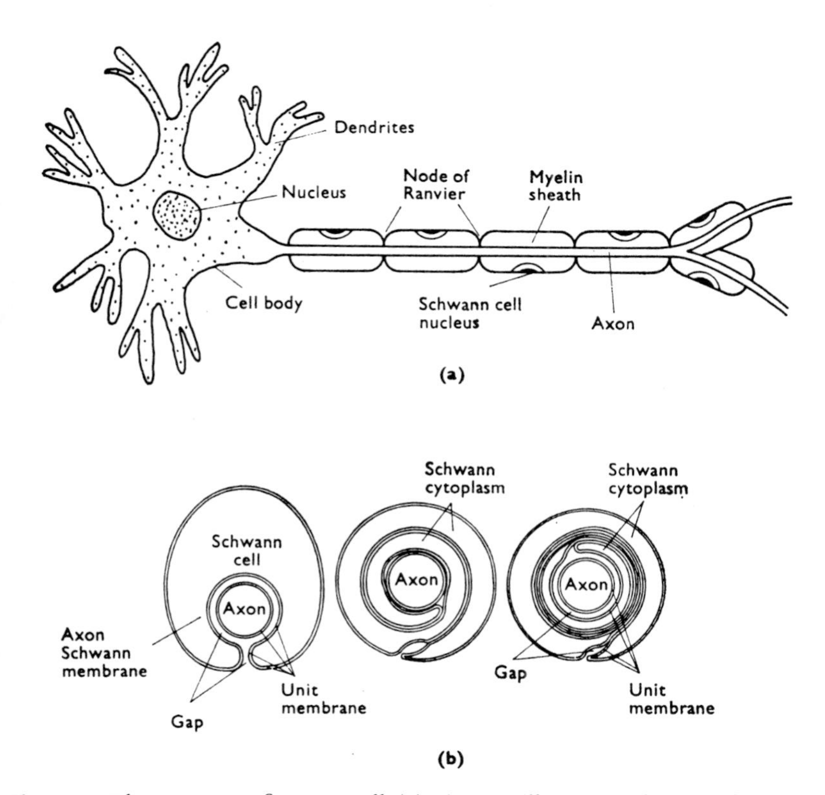

Fig. 8–1 The structure of a nerve cell. (**a**) Diagram illustrating the main features of a myelinated nerve. (**b**) The manner in which a Schwann cell migrates around an axon as the myelin sheath is laid down. ((**b**) from ROBERTSON, J. D. (1966). *Symp. Internat. Soc. Cell Biol.*, 5.)

postganglionic fibres of the autonomic nervous system) are bounded by a *myelin* sheath. This sheath, which is composed of concentric rings of protein and lipid, is formed by non-nervous Schwann cells associated with the axon not by the axon itself. The mode of production of the sheath is remarkable. The Schwann cell rotates around the axon so that successive layers of its cell membrane are laid down one on top of the another. As each layer is added to the one below the cytoplasm is squeezed out so that plasma membranes from opposite sides of the Schwann cell are in contact with one another (Fig. 8–1). The whole process may be likened to the winding of a long balloon tightly around a walking-stick in such a manner that the air is eliminated from between the successive layers of rubber. The multiple layers of cell membrane which surround the axon as a result of the activity of the Schwann cells greatly increases the electrical resistance between the inside of the axon and the surrounding medium.

Non-myelinated fibres also have associated Schwann cells but, since these do not rotate, there is no multiple myelin sheath around these axons.

Nerve impulses are usually initiated in the cell body and then pass along the axon to the nerve termination. Input signals to a nerve cell arrive via the dendrites or by small processes or other nerve cells, *buttons terminaux*, which end on the cell body.

8.2 Functioning

The means by which a nerve impulse is conveyed along an axon is based on a modification of the membrane properties responsible for the origin of the membrane potential (p. 50). In the case of nerve cells in an inactive state the magnitude of this *resting potential* is of the order of 60–90 mV, with the inside of the cell negative to the outside.

When a nerve is stimulated, either naturally or by an electric shock, change occurs in the membrane potential at the point of stimulus. If an artificial stimulus is of small magnitude the potential change across the membrane will be only a few millivolts at the point of stimulus and dies away quickly as it passes further along the nerve because of the relatively leaky nature of the membrane. Greater stimuli, capable of producing an initial potential reduction of about 20 mV, cause the membrane to reach a threshold, the *firing potential* at which a still larger change is initiated. This change is of short duration, only about 1/1000th second, but the current flow associated with it serves to excite the neighbouring region of the membrane with the result that the potential change (*action potential*) sweeps as a self-propagating wave from the point of stimulus to either end of the nerve. Selective changes in the permeability of the membrane to sodium and potassium ions are responsible for the production of the action potential. Initially, the permeability to sodium is suddenly increased allowing sodium ions from outside the cell to penetrate down the electro-chemical gradient. The positive charge carried into the cell by these ions reduces the internal negativity. The precise number of sodium ions which must penetrate to produce a given change in membrane potential depends on the capacitance of the membrane (usually about 1 $\mu F/cm^2$) but it is important to note that the number required is small by comparison with numbers initially in the cell. The passage of a single nerve impulse results in a virtually immeasurable increase in cell sodium concentration (e.g. the sodium uptake by a squid nerve per impulse is estimated to be less than 2.4×10^{12} ions whereas 1 mole contains 6.02×10^{23} molecules).

Immediately after the change in membrane potential reaches its peak during the action potential the sodium permeability of the membrane decreases and an increase in potassium permeability occurs. The resultant escape of positively charged potassium ions from the axon restores the membrane potential to near its normal resting level (Fig. 8–2a).

During the course of the action potential the active region of a nerve cannot be stimulated to further response. The time during which this insensitivity lasts is known as the *absolute refractory period*. For a further period of a millisecond or so after the action potential has passed the membrane remains in a *relative refractory* state in which it can be stimulated to a further response but only by a stimulus of a magnitude greater than that normally required. Large diameter nerve fibres tend to regain the normal state more rapidly than small axons, and as a result, the larger motor nerves of vertebrates can conduct up to about 1000 impulses/sec experimentally though probably in life the rate rarely exceeds 100–150/sec.

In non-myelinated nerves the spread of excitation passes smoothly along the axon (Fig. 8–2b). The larger the diameter of a fibre the more rapidly is the impulse conducted. However, in the case of non-myelinated fibres even such large axons as the 1 mm diameter giant fibres of the squid conduct only at a rate of about 6 m/sec. Much faster rates of conduction are achieved by myelinated fibres, the most rapid known at present being in the giant fibres of the prawn, *Pennaeus japonicus* where the rate is 210 m/sec or about 755 km/h.

Myelinated nerves conduct more rapidly than non-myelinated axons because the presence of the myelin sheath greatly increases the electrical resistance of the membrane. This diminishes the short-circuiting effect of current leakage across the membrane and hence permits the current to pass a greater distance along the axon with relatively undiminished intensity. By this means current engendered by the potential change at one node of Ranvier is enabled to reach and excite the next node (Fig. 8–2). The action potential therefore progresses by a series of saltations (jumps) from node to node and the ultimate rate of passage of the nerve impulse is limited primarily by the time taken to develop the action potential at each successive node.

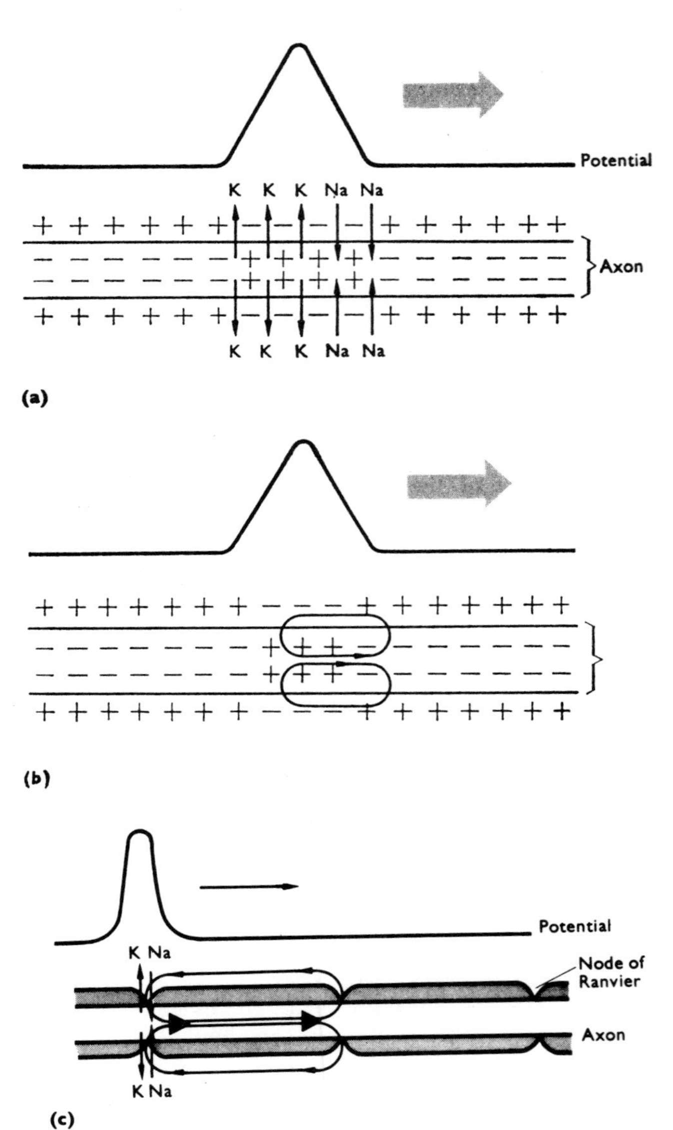

Fig. 8–2 The condition of a nerve impulse. (**a**) Ion movements across the membrane responsible for the production of the action potential. (**b**) Current flow between regions of different potential exciting neighbouring regions of the axon to activity. (**c**) Saltation of current flow between the nodes of Ranvier in a myelinated nerve.

8.3 Release of neural transmitters

In most cases nerve terminations are separated from the effector organs which they stimulate by a distance of about 10–20 nm. Nerve impulses do not cross this gap directly, instead, the arrival of an impulse at the nerve ending triggers the release of a chemical which diffuses across the gap and excites the membrane of the target organ. Various transmitting agents are released by different nerves; the substance utilized by vertebrate motor nerves is acetyl choline (ACh).

Acetyl choline is stored prior to release in small membrane-bounded vesicles near the nerve ending. A little ACh is continuously released in small, relatively uniform, amounts (quanta) but the arrival of a nerve impulse greatly increases the number of quanta liberated. Furthermore, the larger the number of impulses arriving per unit time the greater is the amount of transmitter released through the membrane per impulse.

Specialization of this process has given rise to several neural-hormonal systems: the release of the anti-diuretic hormone (ADH) from the pituitary; the release of adrenalin from the adrenal medulla; the release of hormones from the sinus gland of crustacea, are all examples.

8.4 Electrical transmission from cell to cell

The giant axons which are present in the nerve cords of earthworms and crustacea represent closely conjoined individual nerve cells. The junction between adjacent cells is sometimes lost but the lateral giant fibres of prawns still display septae which represent tight junctions of the plasma membrânes between successive cells. Impulses traverse these septae electrogenically and hence activity in the giant fibres passes each one with a delay of only about one-tenth that at junctions where transmitting agents are involved. Conduction between neighbouring epithelial cells can also occur across tight junctions. In most giant axons impulses moving in either direction pass the septae with equal freedom. However, in at least one somewhat similar nerve–nerve junction impulses can cross in only one direction. Motor axons to the flexor muscle of the abdomen of decapod crustacea (the muscles responsible for the 'tail' flip of lobsters and prawns) make synapses with both the medium and lateral giant fibres in the nerve cord. The motor axon can thus be stimulated by its own cell body or by either of the giant axons. Since the giant fibres make connection with many muscles it would clearly be undesirable, to say the least, for activity in one motor axon to be transmitted to the giant axons themselves. To prevent this happening the synapses between the giant axons and the motor nerves are so modified that impulses can pass from the giant axon to the motor nerve but not in the reverse direction. How the plasma membranes at the junction achieve this power of rectification is as yet unknown.

Further Reading

General

FAWCETT, D. W. (1966). *Atlas of fine structure: the cell, its organelles and inclusions.* W. B. Saunders, London and Philadelphia.

JÄRNEFELT, J. (ed.) (1968). *Regulatory functions of biological membranes.* Elsevier, Amsterdam, London and New York.

Plasma membrane

BRETSCHER, M. S. (1973). Membrane structure: some general principles. *Science*, **181**, 623–9.

NICHOLSON, G. L. (1976). Transmembrane control of the receptors of normal and tumour cells. *Biochim. Biophys. Acta*, **457**.

Golgi apparatus

BEAMS, H. W. and KESSEL, R. G. (1968). *Int. Rev. Cytol.*, **23**, 209–76.

Microtubules

OLMSTEAD, J. B. and BORISY, G. G. (1973). *Microtubules. Annu. Rev. Biochem.*, **42**, 507–40.

WILSON, L. and BRYAN, J. (1974). Biochemical and Pharmacological properties of microtubules. *Advances in Cell and Molecular Biology*, 3, 21–72.

Lysosome system

ALLISON, A. C. (1968). *Interaction of drugs and sub-cellular components.* Churchill, London.

DE DUVE, C. and WATTIAUX, R. (1966). *Ann. Rev. Physiol.*, **28**, 435–92.

History and evolution of membrane systems

NASS, S. (1969). *Int. Rev. Cytol.*, **25**, 55–129.

Lipids and membrane configuration

VAN DEENEN, L. L. M. (1968). 'Membrane lipids' in *Regulatory functions of Biological Membranes.* Edited by J. Järnefelt. Elsevier, Amsterdam, London and New York.

HOKIN, L. E. (1968). Dynamic aspects of Phospholipids during protein secretion. *Int. Rev. Cytol.*, **23**, 187–208.

Cell contact and adhesion

STAEHELIN, L. A. (1974). Intercellular junctions. *Int. Rev. Cytol.*, **39**, 249.

REVEL, J. P. (1974). Contacts and junctions between cells. *Symp. Soc. exp. Biol.*, **28**, 447–61.

Ion and water transport

GUPTA, B. L. (1975). Water movement in cells and tissues. Vol. 1, *Perspectives in Experimental Biology.* Edited by P. Spencer Davies. Pergamon Press, Oxford.

MAETZ, J. (1975). Aspects of adaptation to hypo-osmotic and hyper-osmotic environments. *Biochemical and Biophysical Perspectives in Marine Biology.* Edited by D. C. Mallins and J. R. Sargent. Academic Press, London and New York.